Log Drives on the Connecticut River

Log Drives on the Connecticut River

Bill Gove

Bondcliff Books
LITTLETON, NEW HAMPSHIRE

Log Drives on the Connecticut River

Library of Congress Card Number: 2003106902
ISBN 1-931271-08-9

Text composition by Passumpsic Publishing, St. Johnsbury, Vt.
Printed in the United States by Sherwin Dodge Printers, Littleton, N.H.

Additional copies of this book may be obtained directly from:
Bondcliff Books
P.O. Box 385
Littleton, NH 03561

Contents

Acknowledgments

I wish this story could have been completed much earlier, say 70 or 80 years ago. So much has been lost about the history of this fascinating era, much of which can never be repeated. What I've been able to probe from fading memories and dusty archives only tells part of the story; much of the human element was never recorded.

There has not been a great deal written about the epoch of the river driver. Stewart Holbrook and Robert Pike provided us with some of the romantic side of the rivermen, for which we're grateful. Fortunately the research for this volume was begun many years ago when there were a very few of the old rivermen still with us. I'm grateful for the times I was able to sit and listen to the recollections of Bill James, Tom Cousy, Bill Dexter, Arthur Hughes, Perley Hurd, Harry Marsh, Marv Cook, Dolphe LeClaire, and the widows of Phonse Roby and Robert Craig.

The annual log drives were always an exciting event for the people living along the river and consequently a surprising number of photos were taken of the action. Searches through old albums of the descendents, and at various libraries, revealed some valuable images. I'm especially indebted to Aubrey Janion of Plainfield, New Hampshire, who many years ago made prints available from the glass plate negatives taken by Leonard Sherman on the river drives. I am also gratified to have had access to the large selection of photos preserved by Stone and Webster Co. (the last owners of the Connecticut Valley Lumber Co.) in their 1914 report. The late Robert Pike, who has preserved so much of the folklore of the Connecticut Valley, was kind enough to pass on some information and photographs. Another contributor of photos and firsthand knowledge was the late Lester Fogg of Northumberland, New Hampshire.

The libraries of the Vermont Historical Society and the New Hampshire Historical Society offered a fine source of material, as did the town libraries of Bellows Falls, Vermont, and Northampton, Massachusetts, in particular. We can be grateful for the efforts of the libraries in preserving the old newspapers, a fine source of historical information.

The popular histories of logging have frequently tended to pass on stories which are either fictional in content or exaggerated beyond the limit of credulity, and which are all too frequently accepted as factual. Unfortunately this hyperbole has sometimes presented a misleading image to later generations. I have tried to either eliminate such anecdotes or at least hint at their questionable veracity. The truthful accounts of the colorful rivermen are enough in themselves to illustrate a dramatic and colorful occupation.

Gratitude is also extended to Fred Cowan, former district manager of the St. Regis Co. timberlands, for his time in going over the manuscript and helping to improve accuracy with his knowledge of logging history. And I am very grateful to Diana Gauthier for her many hours spent helping me with the manuscript.

Foreword

The tiny steam-powered scow wedged her way across the sullen waters of the Connecticut River, intent on a mission easily accomplished through years of experience. The wedge-shaped hull pushed through the mass of floating logs in order to make secure a tow of logs that was to have historic significance. The rickety little craft, all lungs and very little body, had endured many years on the river but was still capable of the task now at hand. The gray skies of the late fall day didn't discourage the large group of onlookers intently registering every movement made by the rusty little craft, as if it was a sight never to be beheld again. For a truth, it was probably just that.

Reaching the east bank of the Connecticut, the scow made fast to a boom or string of logs connected endwise by chains. Chugging across the river in a crescent-shaped path to the flat meadows on the west bank of the Connecticut River, the scow tightened the boom which by then had spanned the river and bent with the current as it strained the chains that held the logs together. Cameras clicked and somber faces watched, some in disbelief, as the last boom of logs was brought in to corral what was apparently the last long-log drive ever to be held on the Connecticut River. The year was 1915. The company, the Connecticut Valley Lumber Company, had strongly alluded that there would be no repeat the next year of the renowned massive annual log drives for the company sawmill at Northampton, Massachusetts.

A Connecticut River without log drives was unthinkable to the scores of observers, as well as the residents along the upper part of the river. Standing on the prow of the little watercraft was Al Patrick (Albert Kirkpatrick by birth), one of the company's log drive bosses for the last 32 years, showing no outward signs of the tremors he must have felt within. Lumbermen were not prone toward exhibiting visible evidence of emotions.

On board, in addition to the boat's operator, was another veteran of many years of Connecticut River log drives, Rube Leonard, eager to take a personal hand in booming what he assumed to be the finale to that annual big event, the most exciting event ever on New England's "long river."

Al Patrick had been a crew member on the company's first log drive, back in 1869, and had seen many drives go over the dams since then. Rube Leonard had likewise been a participant in that initial drive of the company and was the only man said to have worked on every log drive since the beginning of the Connecticut Valley Lumber Co.

Their thoughts numbed by disbelief that this could ever happen, bolstered by the rumors that the timber resources still abounded way upriver in the North Country, they must have pondered the future of the thriving lumber industry.

The little scow pulled the boom of logs into the entrance of the oxbow in the river, that circular stretch of water that encircled the island where the large company sawmill was located at Mount Tom, below Northampton. Al Patrick then secured the boom in a closed position. Another winter cut of sawlogs was in place, quite possibly the last one.

Could this be the end of an annual event that the North Country loggers thought would surely go on forever?

Introduction

The Connecticut River has its source in the three Connecticut Lakes, way up in the northern tip of New Hampshire, in a region blanketed with the northern forest of spruce, balsam fir and northern hardwoods. The river appears to begin at Third Connecticut Lake, but a little further north, only a stone's throw from the Canadian border, there's actually a Fourth Connecticut Lake, though hardly large enough to even be called a pond.

By the time the river courses from the Second to the First Connecticut Lake there is just about enough water to float a log, that is, with the ingenious help from the loggers. Fed by scores of streams and small rivers that flow into the Connecticut from the green carpeted hills and mountains in New Hampshire's Coos County, the Connecticut River quickly grows in volume during the first few miles of flow.

Flowing about 440 miles from the Canadian border to the Atlantic Ocean, draining 11,300 square miles of rich forestland and fertile alluvium, the Connecticut River is New England's longest river. Drawing its bulk from four states, the river forms a regional valley that has remained an area of distinct physical and cultural characteristics.

The Connecticut, the "Great River," has been an artery of commerce for many generations of settlers throughout its extensive water course. The river was formerly navigated above tidewater more than any other river in New England. On one unique and noteworthy journey a steam-powered vessel surmounted the physical barriers to steam as far north as Wells River, Vermont. America's first canals and locks were also built on the Connecticut River at Bellows Falls, Vermont.

Not the least of the river's laurels is the distinction of hosting what some say was the longest log drive in the nation. The possibility exists that the Mississippi River may have floated a longer drive in miles, but none compared with the Connecticut River when considering the volume of logs involved. Huge masses of long spruce logs were floated and goaded down the Connecticut for many generations.

The most abundant natural product of the region drained by the Connecticut River, the product that sustained the residents for many generations, was the rich forest resource. The profuse forest growth provided the resource for a flourishing lumber and paper industry that was able to furnish employment for many decades ahead. From the very beginning of settlement until the early years of the twentieth century, the Connecticut River served as an avenue of transport to move voluminous quantities of sawlogs and pulpwood, culminating in the massive log drives of the late 1800s.

The story of these log drives is the saga of rugged, fearless loggers and rivermen who were so devoted to their livelihood that they seemed to live solely in anticipation of the next spring's log drive. They faced death daily on the drive, with nearly every step their spiked boots took; they saw some of their workmates drown before them every year. Yet they were there the next spring at ice-out, ready to apply their acquired skills and colorful riverman's lingo. The skills are gone and so is much of the history, locked in the minds of loggers long gone. But a little of that history remains, and the story is one of an era that will never be repeated.

PART I

1730–1860

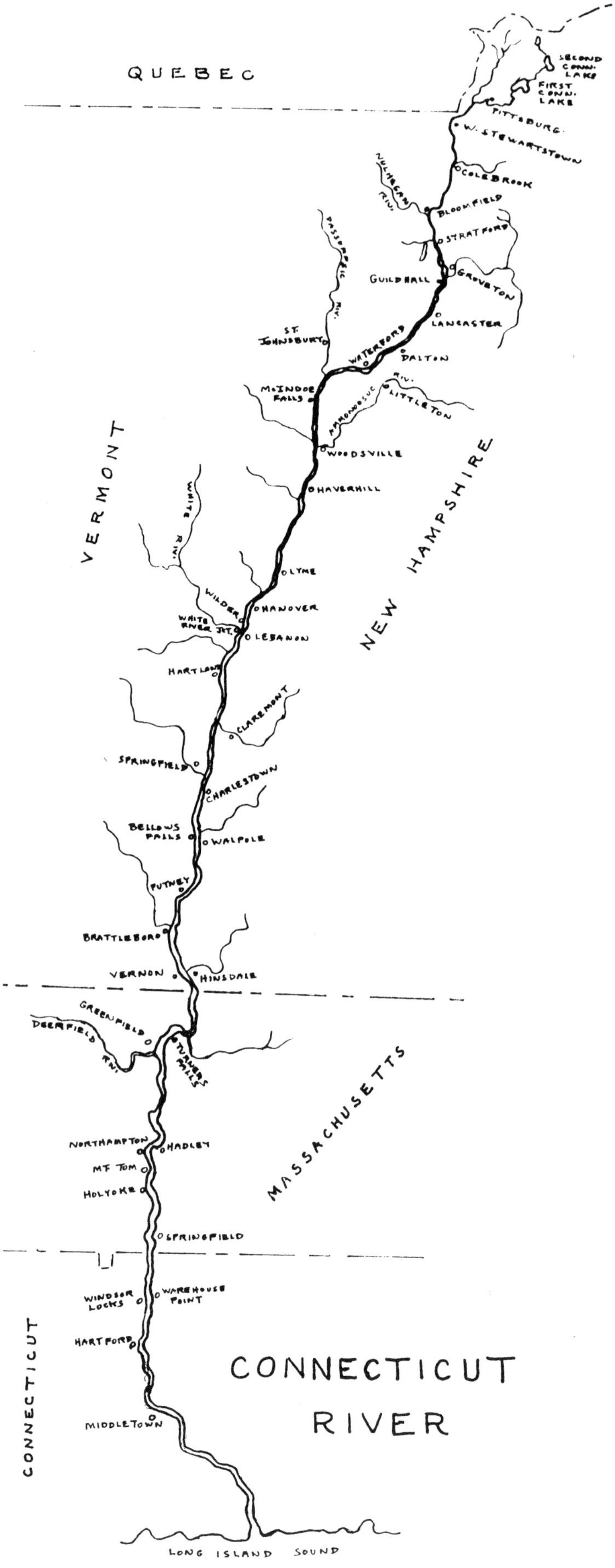
QUEBEC
SECOND CONN. LAKE
FIRST CONN. LAKE
PITTSBURG
W. STEWARTSTOWN
NULHEGAN RIV.
COLEBROOK
BLOOMFIELD
STRATFORD
PASSUMPSIC RIV.
GUILDHALL
GROVETON
LANCASTER
ST. JOHNSBURY
WATERFORD
DALTON
McINDOE FALLS
AMMONOOSUC RIV.
LITTLETON
WOODSVILLE
VERMONT
HAVERHILL
WHITE RIV.
NEW HAMPSHIRE
LYME
WILDER
HANOVER
WHITE RIVER JCT.
LEBANON
HARTLAND
CLAREMONT
SPRINGFIELD
CHARLESTOWN
BELLOWS FALLS
WALPOLE
PUTNEY
BRATTLEBORO
VERNON
HINSDALE
GREENFIELD
DEERFIELD RIV.
TURNERS FALLS
MASSACHUSETTS
NORTHAMPTON
HADLEY
MT TOM
HOLYOKE
SPRINGFIELD
WINDSOR LOCKS
WAREHOUSE POINT
HARTFORD
CONNECTICUT
MIDDLETOWN
CONNECTICUT RIVER
LONG ISLAND SOUND

1

The Early Log Drives

THE STORY OF log drives on the Connecticut River actually had its beginning in the Colonial era, possibly around 1730. Lumbering activities on the Connecticut in the eighteenth century were not large, and thus log drives on the river were relatively small in volume, especially when compared with what was to come down the river 150 years later.

Until the mid-nineteenth century there were two distinctly different ways of handling logs floating down the Connecticut River —rafting and loose log drives. With rafting the logs were fastened and held together in a boxed raft for the entire journey. Loose log drives allowed the logs to travel unattached and move as the current urged them. The more common method used for the longer drives during this early era was rafting.

LOOSE LOG RIVER DRIVES

Records of the movement of logs on the Connecticut River show that as early as 1732 there were sawlogs floated down the river to provide lumber for the English navy. These sawlogs were often mixed in with the movement of the long mast logs on the river. Until the late 1700s there were no man-made obstructions on the river and only a few natural ones to impede the free movement of the logs.

Logs floating loose, however, could be quite damaging to riverbanks and choice riverside farmlands. This was a problem the rivermen would have to contend with from the beginning of the log drives until the day that they ceased.

Even prior to 1800, for example, lawsuits were filed in Massachusetts because of logs constantly being lodged on Captain Gunn's Island in the town of Montague, Massachusetts, forcing the town to maintain a yearly inventory of "masts and mill logs" deposited on the island each spring. The State of New Hampshire passed legislation in 1792 that required that any logs or masts stranded on land must be removed by the 30th of May or be forfeited to the landowner. This same legislation allowed reimbursement of triple value for any timber stolen from the river, a frequent practice. Then, just prior to 1800, man-made barriers began to appear in the river, intensifying a conflict during an age when there were no environmental laws to control man's use of the river.

The conflicts in river use between the land and the rivermen at this early date were well exemplified by a New Hampshire legislative act in 1814 regarding property liability on the river. The legislation authorized Walpole resident Josiah Bellows II to float his pine timber down the Connecticut River between April 20 and December 1. In so doing, however, he was required to put the logs in the river below the dam and canal at Bellows Falls in order to prevent problems at the canal. Being aware of the damages and liabilities encountered with log driving, the lawmakers required that Bellows post $20,000 in liability bonds with the owners of the down-river bridges at Westmoreland and Hinsdale, New Hampshire. Any logs deposited on land adjoining the river became property of the landowner if not removed within 10 days, and Bellows was liable for damages.

The Westmoreland bridge took quite a shaking eight years later, however, from logs being driven by Timothy Shedd of Newbury, Vermont. Shedd was a tanner by trade who also owned a sawmill on Vermont's Wells River. He had employed an experienced river driver, Hardy Chamberlain, to drive about 100 long mast logs and other timbers from Barnet, Vermont, south to Hartford, Connecticut, beginning in April 1822. By that date, however, there were many new problems. Canal locks had been built on the river in five different locations and dams spanned the river at the canal heads. Many of the large logs were too long to go through the various locks and had to be coaxed over the dam and falls. At the Quechee locks, Chamberlain was able to pry the long logs through the locks one at a time with an iron bar. But at Bellows Falls he had to turn them loose over the falls singly, with some being caught and lodged on the rock face of the falls. About 24 sticks got hung up, eight of which were never retrieved.

Reaching the bridge spanning the river at Westmoreland, Chamberlain's crew took on an even larger bundle of trouble. A mid-July freshet had raised the water level in the river and brought his logs hard against the wooden bridge support piers in the river, knocking out three of them and leaving $400 in damage. Shedd was sued for damages by the Westmoreland Bridge Company based on the contention that the "flood wood" (loose logs and other debris) would not have done the damage without Shedd's long timbers mixed with it. However, Shedd was judged by the court to be not guilty of negligently handling his timber, because the newly installed piers were placed only 20 feet apart and were thus ruled to have impeded river navigation.

PINE MAST LOGS

An important and valuable component of these early loose log drives was the mast log, long sticks used for construction of the masts and spars on sailing ships. Although the shipbuilding industry on the Connecticut River was never able to grow to the proportion found on such rivers as the Kennebec or the Penobscot in Maine, it did provide a market for the plentiful supply of red oak, white oak, and white pine timber along the lower Connecticut River. The primary market for mast logs in the early years, however, was for export to England, and eventually to France after the American Revolution.

The English Navy reigned supreme in the sixteenth and seventeenth centuries, but in that era the success of navies rose or fell with the availability of mast material. Dockyards had to have the ability to fit or replace masts that could withstand the crushing force of sail carried in storm or in battle. England had leveled most of its thick forests of hardwoods and conifers as early as the thirteenth century. Over the next several centuries, the preserved royal forests were also destroyed. Expansion of shipbuilding and ocean commerce at the beginning of the seventeenth century forced England to look elsewhere for mast wood. Forests on the shores of the Baltic Sea were then turned to as an available, though costly, source of mast material.

Scots pine (Riga fir) from the Baltic countries made excellent mast material and hull timbers. Transportation costs were excessive, however, added to by the political and military aspects of keeping open the trade routes. The importance of these trade routes underlay British Naval policy for many centuries, but it eventually became apparent that her new colony in America must become a secondary source of mast material, despite the distance.

A shipload of pine masts arrived in England from Jamestown, Virginia, in 1609, and the first load from the New England shore was shipped from Penobscot Bay in 1634. The large white pine masts from New England were welcomed by the shipwrights, who by this time were banding together smaller sticks of Baltic fir to fabricate a tall mast. However, the Admiralty was slow to develop the American market as a solution to their timber crisis, and it wasn't until after 1652 that an annual trade with New England was established. By this date, there were colonial settlements willing to provide any type of timber for an available market.

The white pine from North America was found by the English shipwrights to be slightly inferior to the Baltic wood as regards to strength, but it was lighter in weight and easier to work. Another advantage of the white pine was the large tree sizes then available from the shores of New England, large enough that a single stick could make a lower mast.

The felling and log movement when harvesting a pine mast log required special care and techniques. White pine is not a strong wood, and the large trees were massive and heavy, breaking easily if not handled with care. Techniques used were quite unique.

A ground bed had to be prepared on which to fall the tree, but even then the valuable main trunk would sometimes shatter beyond usefulness. Preparing a bed meant removing obstructions in the direction of the tree fall and then cutting smaller trees and laying them down in such a way that their branches made a bed to help break the fall. Snow made the best bed, of course. Making these ground beds, though time-consuming, was a necessity, for the masts were a very valuable commodity.

The length that the mast stick was cut from the fallen tree would depend on the size and quality of the tree. The general rule was to allow three feet in length for each inch in tree diameter at the base. Thus a tree 40 inches at the base could yield a 120-foot long mast log, and such lengths were common. A large lower mast required for an eighteenth-century first-rate warship with 100 guns had to be 120 feet in length with at least a 36-inch diameter at the base. The bowsprit of such a warship typically measured 38 inches at the base, and 75 feet in length. Even though the principal masts frequently had to be fabricated or "coaked" together from squared timbers, as much length as possible was taken from the great pines. Value obviously increased with greater length; a 32-inch diameter mast hewn to proper shape was worth 56 pounds, while a 34-inch mast was valued at 90 pounds currency.

Records of the old growth white pine once existing in the original New England forests reveal the large size they had attained. For example, a pine in Dunstable, Massachusetts, measured 7 feet 8 inches in diameter when harvested in 1736, and a pine in Lancaster, Massachusetts, stood 264 feet tall in 1817. And in Hopkinton, New Hampshire, it took 55 yoke of oxen to draw a huge pine log cut there. Unfortunately, the log later broke while passing over Amoskeag Falls in nearby Manchester. The tree left a stump large enough in diameter for the logger to turn a yoke of oxen around on it.

Drawing the long mast log from the stump to the river location became a very specialized operation. A reasonably straight skid road was essential, of course, with only gradual curves allowed. It was no easy task to "balk" or haul the long mast stick without damage resulting. Roadways cut on a side hill had to be "wharfed" or built up with log material to make a level road. If the log was to be drawn on wheels instead of sledded, each log end was raised by levers and hung by strong chains under a wheel axle. The wheels must have been huge; New Hampshire historian Jeremy Belknap wrote that wheels 15 to 18 feet in diameter were used in logging the old-growth pine.

A later adaptation of the wheeled conveyance used smaller diameter wheels and mounted the mast on top of the axle. Getting the mast on the axle top involved a maneuver that was quite skillful. The entire wheel assembly would be tipped up on one end with the axle vertical and the mast then rolled over on top of the bottom wheel. After chaining the stick firmly to the upright axle, the whole assembly was jerked back upright again by oxen power, with the log resting on top of the axle. A pair of wheels might be placed on each end of the mast stick.

If the mast was being drawn out on snow, the task was obviously somewhat easier. The front end of the stick was placed on a sled and the rear end allowed to slide in the snow. Originally the large or butt end of the log was put on the cross bar of the sled in order to take more of the weight off of the drag. However, problems were encountered on side hills, where sometimes the mast would roll and upset the sled. This problem was frequently avoided by placing the small end of the log on the sled and fastening it to the tongue of the sled by a swivel chain. Long sticks might have two or more sleds under them.

Whether moved by wheels or sled, the motive power was oxen, with 8 to 20 pairs yoked together in tandem in front of the load. Goading the oxen to move together to start the ponderous load was a difficult task—called "raising them" by the loggers—and it taxed all of the teamsters' skill and profanity. When descending a steep hill the load would have to be held back to prevent it from pushing or "sluicing" the oxen down the slope. A yoke or two of oxen might be attached to the rear of the mast to act as a brake, a practice called "tailing."

As the long log rose over the crest of a hill or spanned a narrow depression, the oxen were sometimes left in a precarious position, actually hung in the air. The sled or wheels to which they were attached might be tipped off of the ground, even suspending the oxen while fastened in their yokes. Then, as the log moved forward, powered by the other oxen which still had their feet on the ground, the suspended animals would drop to earth, possibly injuring or even killing them. Rather than stop the movement of the massive log, injured or dead animals were cut out and replaced on the move by spare oxen kept along the route of travel.

Getting the long log into the waters of the Connecticut River took many strong hands, but at this point a new and different journey began in order to reach the seaport down the river. It was a journey fraught with a host of problems as the excessively long stick had to be guided over the many rapids and falls. Occasional breakage also occurred in this phase of the operation. On reaching the seaport at locations such as Middletown, Connecticut, the masts were loaded onto English cargo ships especially designed to carry masts as cargo. The sticks were loaded end-wise through oversize ports in the ship's stern. These cargo ships carried a crew of about 25 and loaded a burden of 400 to 600 tons, and in some cases up to 1,000 tons. That meant a cargo of from 40 to 100 mast sticks, plus the smaller sticks such as spars, bowsprits and yards in each load.

Most of these white pine masts from the Connecticut River were harvested in the Connecticut Valley above Northampton, Massachusetts and on up into New Hampshire and Vermont. Some pine grew in the Connecticut colony also, but oak was the principal wood exported from the lower river. Masts were an important export product on the river for many years, but the Connecticut River valley never did become a major mast port such as did Portsmouth, New Hampshire, or Falmouth, Maine.

It wasn't many years before the accessible large white pine suitable for masts began to be sparse in number throughout all of New England. England's wars with Spain and then the Dutch in the mid-1600s, followed by an impending war with France in the latter part of the century, resulted in stepped-up naval construction and a resulting timber crisis. Added to that was the newly flourishing lumber market for the colonial sawmills. Therefore, as early as 1685 the British Admiralty felt that strong measures were needed to protect the suitable mast trees yet remaining. Surveyors of pines and timber were appointed by the Crown to oversee the mother country's mast interests in New England. Any suitable mast pines within approximately 10 miles of any usable waterway were blazoned with the king's mark for preservation and exclusive use by the Crown. This legislative action became collectively known as the Broad Arrow Policy.

The identifying mark put on the chosen mast trees was the ancient mark emblazoned on all property of the Royal Navy, what became known as the "broad arrow." The symbol was a three-legged letter "A" without the horizontal bar. Pine trees with this "broad arrow" still distinguishable were to be found along the Connecticut River over a century later.

This despised policy was enacted piecemeal from 1690 until 1729 and was eventually felt throughout the New England colonies. An Act of May 10, 1708, preserved all mast trees larger than 24 inches at a point 12 inches above the ground within Her Majesty's Province of New Hampshire. Later acts tightened the law to include smaller trees. When the original charter was granted for the Connecticut River town of Haverhill, New Hampshire, it included the provision that "all white pine and other pine trees within the said township fit for masting our (the King's) royal navy may be carefully preserved for that use, and [are] not to be felled without our (the King's) special license for so doing."

The Broad Arrow Policy continued to affect the timber industry on the Connecticut River until the outbreak of the American Revolution in 1775.

Reaching farther north on the Connecticut River for additional sources of mast trees, free-roving gangs of cutters traveled into the upper valley during the early 1700s. There were no settlers to interfere with the cutting at that time, because prior to the 1750s there were no settlements above the Massachusetts boundary line, except for a few settlers at Fort Number Four in Charlestown, New Hampshire. Only after Roger's Rangers had destroyed the Abenaki Indian village of St. Francis near the St. Lawrence River in 1759, followed by the end of the French and Indian Wars in 1763, did the settlers feel encouraged to travel northward in search of new home sites.

Mast cutters would travel north in canoes loaded with supplies and set up camp where large pines could be felled in proximity to the river. It was stated by historian Lyman S. Hayes that there were records of masts put in the river near Rockingham, Vermont, as early as 1732. The first settlers in Cornish, New Hampshire, are also said to have found a mast camp near the mouth of Blow-Me-Down Brook. Another former mast camp was known to have been located just below the old covered bridge in Windsor, Vermont.

Maine's Piscataqua River and New Hampshire's Merrimack River had maintained a monopoly on the mast trade for the Royal Navy for many years. However, when the shortages in southeastern New Hampshire became apparent in the mid-1700s, Connecticut made an effort, though unsuccessful, to become a serious contender. The man behind that ill-fated effort was a colonial agent from Connecticut, Jared Ingersoll, who sought the Navy Board's interest in a fresh supply of masts from the Connecticut River. Mast pines, he declared, were plentiful above Northampton.

In 1761, Ingersoll was given a contract by the Navy Board to procure 80 mast pines, enough for two shiploads, on the Connecticut

River. These were to be cut between Deerfield, Massachusetts and "ye Cohees" or Coos, what is now Haverhill, New Hampshire. Ingersoll soon put a logging contractor to work, who proceeded to encounter just about everything that could possibly go wrong.

He felled 89 white pine suitable for masts, five of which broke when they fell and were thus worthless. Also cut were another 63 "defective" trees and these 147 sticks were "balked" to the river with his oxen teams. Two more of the large mast sticks broke when rolling them down the bank into the river. Another fine 30-inch mast had to be abandoned because it still lay quite a distance from the river when a five- to six-day supply of hay for the oxen was stolen by a disgruntled resident.

The drive down the river proved to be a difficult one too, with the long logs hampered by low water. Twenty-one logs jammed tight at one of the rapids and couldn't be freed until the following spring. Another of the large 36-inch mast sticks broke going over the Bellows Falls; another one was damaged at the Deerfield rapids. By the time the logs reached the lower Connecticut there were fewer than 80 mast sticks and the contractor had to go back for more.

Ingersoll's troubles weren't over yet, either. While the logs were awaiting the cargo ship in Middletown, Connecticut, Governor Benning Wentworth sent a deputy, Colonel Symmes, to seize all the logs that were shorter than the contract length. Wentworth supposedly held a monopoly on the mast trade in New Hampshire and Massachusetts, and it was his intended action to forestall and interfere with the shipment. As a result, Ingersoll's masts didn't reach England until 1764.

Colony residents chafed under the restrictions of the Broad Arrow Policy imposed by their British masters. They had strong feelings about being denied the use of their available timber resource; resistance was inevitable. Small sawmills had been established early in many of the settlements, and lumber was in demand, of course. Sawing logs into lumber was accomplished at that time either by pit sawing with brute manpower or by an up-and-down saw blade powered by a water wheel. In either case the saw employed was a long vertical one capable of sawing through pine logs as large as 3 to 4 feet in diameter. White pine was the chosen house construction material during that era, and the large clear butt logs gave the saw millers the select grade of lumber they desired to use. Therefore these large mast logs were the ones they wanted for themselves.

Indicative of the superior quality of these great pines is the description given of one of the trees by historian J. Q. Bittinger in his 1888 *History of Haverhill, New Hampshire.* The large tree was growing on the plains near Brian Hill, along a brook above the

Swazey mill pond, and was felled with an axe. The logger removed 70 lineal feet of the tree in the form of logs, the tree measuring four feet in diameter on the butt and two feet at the 70-foot mark. The mill cut 4,000 board feet of lumber, 2,500 as "clear stuff."

The colonists, aware that there was far more profit in their sawing the log for the local market than there was in selling the tree as a mast stick, despised the governmental restrictions on what trees they could legally cut, and did their best to circumvent the King's orders and conceal their deeds. The colonial sawmillers were careful to saw no boards wider than 24 inches, thus destroying any evidence that the boards came from a large tree, and the broad arrow mark on the bark was quickly obliterated.

Catching the colonists in the act was difficult with the limited number of agents available. There were times, however, when the King's men appeared before the sawmillers could finish their deceptive acts. In 1769 a man named Dean of Windsor, Vermont, was arrested shortly after felling 16 mast trees. In defense, he claimed he was clearing a new meadow, but it was shown that he was actually landing the logs on the ice on the Connecticut River preparatory to rafting them in the spring. Further investigation established that he'd also contracted with a sawmiller to cut and raft a half-million board feet of logs at $1 per thousand feet. Dean spent four months in prison in addition to being levied a fine of 800 pounds.

In another instance, in the winter of 1771–72, an operator named Wheelock cut 1500 pine logs in Hanover, New Hampshire, near the Connecticut River. They were to be driven down the river to a sawmill market in Springfield, Massachusetts, but were seized by an agent of the Surveyor General of Massachusetts. Wheelock's assessed damages proved to be greater than the value of the logs. In 1772 there also occurred what was dubbed the "the Pine Tree Riot" in Weare, New Hampshire, during which the sheriff was attacked and beaten by rioters.

There were, of course, many mast cutters loyal to the King's cause who provided a moderate flow of mast logs from the Connecticut River. They were not free of opposition, however. Instead of the scrutiny of the King's agents they suffered the wrath of the colonists. On one occasion, about 1760, a camp of cutters on the river somewhere near Northampton was raided by area settlers. As all their food was stolen, the cutters were prevented from finishing the operation and removing their mast sticks.

As suitable white pine near the Connecticut River became scarcer, the government sent agents further north on the river to emblaze more choice pine with the King's mark. About 1771 John Wentworth surveyed up through New Hampshire to the "White River Falls" (at Wilder, Vermont) on the Connecticut and on up the river to the 45th latitude. In so doing he marked out a number of pine areas adjacent to the upper Connecticut.

However, the claimed shortages of mast trees, plus the obstructionist activities of the Connecticut Valley settlers, were having an effect on England's policies. The Navy Board decreed that the Connecticut River pines were somehow "inferior" to the New Hampshire masts that were being loaded at Portsmouth. The powerful Benning Wentworth is also said to have played his political strategy of preventing the development of a Connecticut River supply in order to favor his vast New Hampshire holdings. At any rate, the Connecticut River watershed never became a major supplier of mast sticks.

The Revolutionary War brought an end to the cutting restrictions imposed by the King and to government control of all masts for export. Trees then became the property of the landholder. The last pre-Revolutionary War shipment of masts to England is said to have reached Britain's shores in July 1775. After the Revolution, there still were a few mast logs from the Connecticut River shipped out by entrepreneurs to available buyers from both England and France. It has been recorded that in 1785 a Henry Porter of Northampton offered to supply annually one or more cargos of masts 34 inches to 39 inches in diameter, driven down the Connecticut River to Lyme, Connecticut.

It was about this time, or possibly around 1800, that a choice stand of pine trees on the lower plain of Bradford, Vermont, was harvested to make masts for the French navy. Landowners Col. John Baron and General Morey contracted to have the long logs floated down the Connecticut River to Middletown, Connecticut, where they were loaded on ships for France. Denoting the high value of a good mast log at that time is the report that one of the mast logs, measuring 40 inches in diameter and 116 feet in length, was valued at almost $20. The river crew must have had a difficult task maneuvering that log over the many falls and rapids en route.

According to historian Frederick P. Wells, pine trees cut in Newbury, Vermont, supplied the material for some of the masts of the French navy during the wars of Napoleon.

The early years of the nineteenth century saw a reduction in the mast trade activity on the Connecticut River, although some mast sticks continued to be floated down the river until almost 1850. By the early 1800s it had become difficult to locate a sufficient number of large white pine accessible to the Connecticut River, and furthermore, shipbuilding had never developed into a large industry on the banks of the lower Connecticut River.

2

The Years of Log Rafting

ALTHOUGH THE PRACTICE of driving loose logs down the Connecticut River was used to a certain extent in colonial days, especially with the transport of mast logs, it was a method that did not gain wide practice in the early 1800s. The old custom of log rafting—the practice of fastening logs or sawn timbers together in a rigid box or unit for floating down the river—became the acceptable way to move forest products down the river.

Loose log driving required a large crew of men and presented the frequent problems of a scattered, even stolen shipment; log rafting employed only a few hands in transit and kept the goods under control. Rafts also provided a vehicle for on-board transport of other merchandise to market. Rafting had its disadvantages, however, the chief being the safe passage through rapids and waterfalls. Many rafts were broken up on the rapids such as those at South Hadley, Massachusetts, or Enfield, Connecticut. Rafting also became impractical when dealing with the massive log drives of later years.

Log rafting on the Connecticut River had its beginnings in the early years of river settlement. Logs provided an inexpensive material with which to build a river craft capable of carrying commercial cargo such as clapboards, shingles or potash, insecure as it often was in rough waters. Early settler John Pynchon sent small rafts of lumber sawed at his Massachusetts sawmills down over Enfield Falls to Connecticut locations as early as the latter 1600s. Claims have been made that board rafts were started below Bellows Falls as early as 1735. If so, it had to be a risky venture at the time. Activity by settlers in the Connecticut Valley was quite limited

The village of Bellows Falls had a path available for boats that were too large to pass through the canal. To get the boats to the other end of the canal, oxen were hired to haul the boat across land. Courtesy of National Life Insurance Co.

above Northfield, Massachusetts, prior to the completion of the French and Indian Wars. The Indians maintained an intimidating presence, and upper river traffic was restricted. After the close of the last of these wars in 1763, however, the rafting of forest products on the upper Connecticut River became commonplace.

Ever since the first Connecticut River voyage by Adrian Block in 1614, when he was stopped by the rapids at Enfield, the natural obstructions in the river had had a limiting effect on commercial traffic. Rafts and flat boats were able to negotiate most of the rapids on a down-river trip along the lower Connecticut River, but traveling back up on the river with a flat boat was a grueling journey. It required brute force with push poles against the unrelenting currents. And further up river through Vermont and New Hampshire there were rapid drops that couldn't be surmounted by any conveyance traveling up-river.

As for rafts headed down the Connecticut River, any commercial products on board would usually have to be unloaded at the head of the rapids, carried around the rough water, and loaded again after the raft hopefully made it down through the rapids. The destructive falls at Bellows Falls necessitated a well-traveled transfer road (through what is now the village location of Bellows Falls) to carry the products to the foot of the rapids. The result was a lively commerce for local oxen teams and wagons. Most rafts were not used for a return trip up river, being dismantled instead at the down river destinations. The logs or lumber comprising the body of the raft could be sold locally to a sawmill or lumber dealer at the docks. Rafts and flat boats used to convey cargo up the river before the construction of the canals would operate in relays between the impassable rapids.

Because of the harm that loose logs would do to mills and bridges and the damages frequently done to farm meadows and intervales when logs became stranded during high water in the river, the New Hampshire legislature passed a regulation in 1808 to encourage rafting. "An Act Regulating the Mode of Putting Pine Timber into the Connecticut River" stated that all pine logs not rafted or under the immediate control of someone and that all logs on the banks which hadn't been rafted shall be forfeited to the person who takes it up. The lumberman would have 90 days to reclaim logs that escaped from a broken raft, however. A similar legislative act in 1826 allowed the landowner to detain logs on improved land until reimbursed for damages. Also, anyone con-

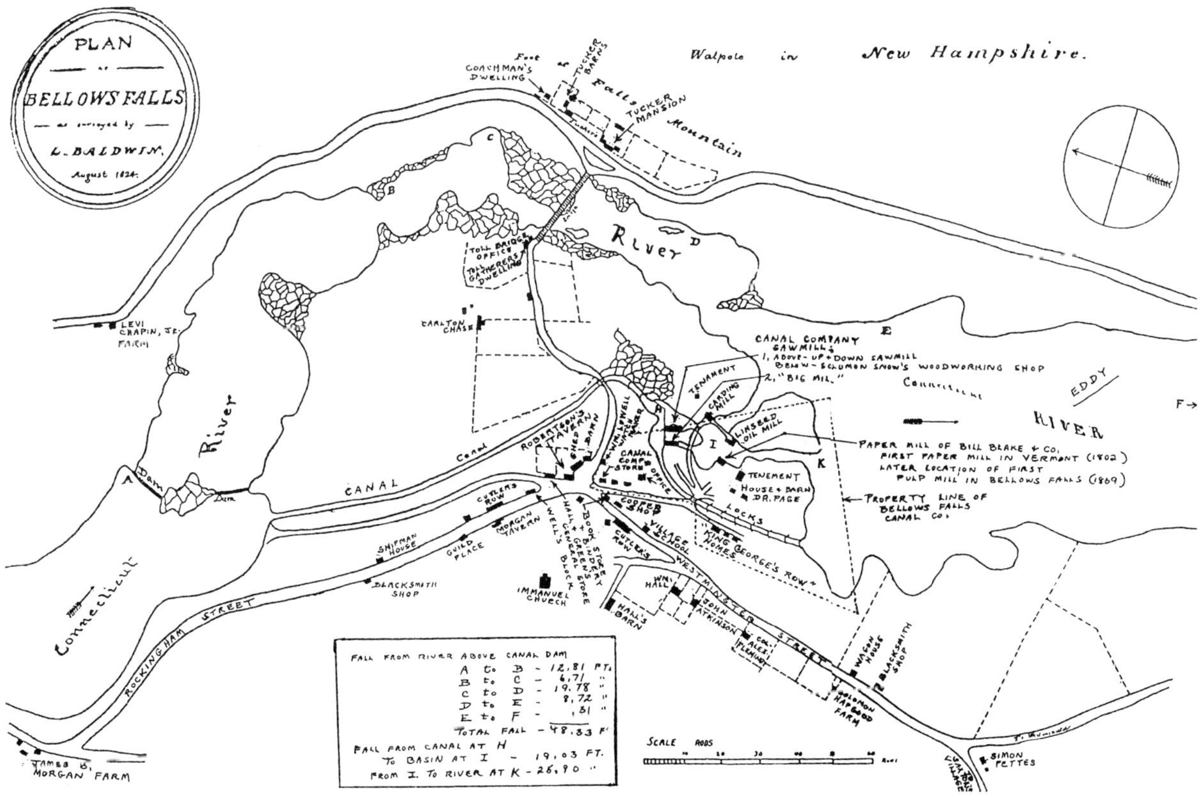

fining or destroying marks on logs or masts would pay their rightful owner six times the value of the logs.

As traffic increased on the Connecticut River with a growing dependence on the river as an artery of commerce, the need for river improvement became evident. During the fifteen-year period of 1795 to 1810 commercial canals were built at five different locations on the river where falls or difficult rapids presented a navigational barrier. Two were in Massachusetts and three further up along the Vermont-New Hampshire border. A sixth canal was later built in Connecticut. This canal system created a practical transportation route for boats as far north as Wells River, Vermont, and was used up until the era of the steam railroads.

These canal installations involved a dam spanning the river and a narrow channel or canal along one side of the river that would regulate the flow of water. The boat or raft was floated into an enclosed lock within the canal and lowered or raised by means of manually operated gates at each end of the lock system, which would drain or raise the water level within the lock. The canal locks were narrow and any flat boat or steamboat using the river had to be designed with the canal width in consideration. This was also the case with the uniquely built Connecticut River log rafts used to supply logs to the mills in Massachusetts and Connecticut.

CONSTRUCTION OF THE RAFT

Early log rafts were constricted in size only by what the crew could handle. Construction of the canals at the beginning of the nineteenth century, however, forced a radical change in raft design. The large rafts were then built in such a way that they could be disassembled into smaller sections for passage through the narrow canal and locks, and then joined together again into a large mass at the foot of the canal. The system worked well.

The logs used to construct a raft were generally 60 feet in length, shorter ones seldom being used. They were fastened together into a rigid unit by placing a 2-inch thick plank across each end of the unit and one across the middle. Holes were drilled through the planks and the logs to accommodate a wooden pin that was driven through the plank and into the log, firmly binding the group of logs together. This solid unit would thus be 60 feet long and about 12 feet wide, narrow enough to fit into the canal locks. The unit was called, in river parlance, a "box." The massive raft itself, however, was made up of many "boxes."

The usual procedure was to join six boxes together to make the raft, forming a tight mass of logs about 180 feet long and 24 feet wide, called a "division." Even larger rafts were occasionally seen

The raft "box" was, at most, 12 feet wide, and reached lengths of possibly 60 feet, narrow enough to fit through the canal lock.

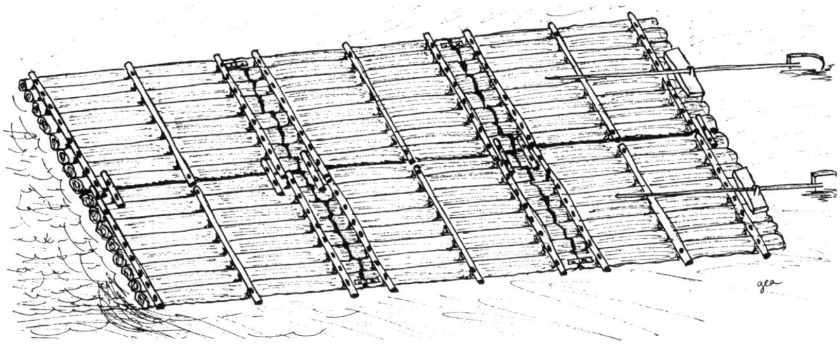

The log raft was usually made of several sections or "boxes" fastened together into a "division" which could be as long as 180 feet.

on the river. Raft dimensions varied somewhat because the rivermen built different sized boxes, but none could be wider than the canal widths. The original charter for the Bellows Falls canal required the locks only be 40 feet long and 16 feet wide. But after the canal was repaired in 1811 it was able to accommodate raft boxes and boats as large as 54 feet by 18 feet with a 2-foot draft.

To assemble the complete raft or division, the boxes were secured together at their corners. The corner pins that held the planks placed across the end of the boxes were left sticking up a foot or more, and short planks with holes were placed over the pins to join the corners of adjacent boxes, thus yoking the boxes together in a flexible arrangement. The tall corner pins would also be used to brace the oars used to steer the rafts. There would usually be crude oars at opposite corners in the back, each manned by one person.

It was not uncommon for gangs of rafts to travel together and share the services of a pilot and cook, even though the rafts might involve many owners. Horace Allbee of Fairlee, Vermont, reported he made a trip down the river in April of 1827 with a gang of rafts totaling 76 individual boxes.

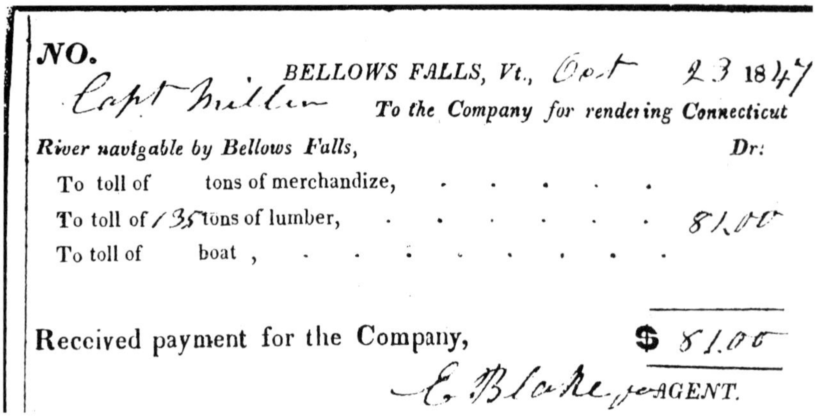
NO.

BELLOWS FALLS, Vt., Oct 23 1847

Capt Miller To the Company for rendering Connecticut River navigable by Bellows Falls, Dr:

To toll of tons of merchandize, ...

To toll of 135 tons of lumber, ... 81.00

To toll of boat , ...

Received payment for the Company, $ 81.00

E. Blake, for AGENT.

Pictured here is an 1847 toll receipt for lumber loaded on rafts or boats passing through Vermont canals.

Goods carried on the raft would obviously have to be those that would not be damaged by occasional wetting: miscellaneous forest products such as clapboards, shingles, and potash, soapstone, slate and various farm produce. Upon reaching the destination, usually Springfield, Massachusetts, or Hartford, Connecticut, the raft would be dismantled and the logs sold to a sawmill.

Rafts were sometimes made up of sawn timbers instead of logs. In the 1830s Nelson Paul was rafting white ash plank that had been sawn at Dunshee's mill on Great Brook in Walpole, New Hampshire. There was a landing on the river not far from the mouth of Cold River where rafts were assembled and top loaded. Various items of lumber and millwork such as shooks of staves were secured on top of the raft. A "stave shook" was a bundle of staves and headings used to manufacture hogsheads or barrels.

William Nelson of Ryegate, Vermont, was one of the earliest lumber manufacturers to ship sawn lumber down the upper river. In 1790 he built a wing dam, sawmill and grist mill on the Connecticut River at East Ryegate Falls, then known as Dodges Falls. A charter was obtained in 1816 to construct a canal and locks around the falls on the New Hampshire side, but it was never built. However, in 1843 a slip was built on the New Hampshire end of the dam to run the lumber rafts over the top of the dam, which by this time extended the full width of the river. This is the same site on the river where in 1906 the Ryegate Paper Co. mill was built.

Below Woodsville there was a large eddy in the river known as Ingalls Eddy where a large boat and raft landing was located. Logs would often be run singly through the Woodsville Narrows above and the rafts formed at this location.

Operating a Log Raft

Negotiating a 60-foot to 200-foot long raft around the bends of the Connecticut River required the skills of an experienced riverman. The crew usually consisted of a pilot and three or four crewmen, two of whom manned the rudder oars. Wages for a pilot were a respectable $1.50 per day. A crude shanty would usually be built on the raft to serve as a kitchen with a box of sand upon which the fire would be built.

Approaching one of the canals the raftsmen would have to swing the raft to shore and bring it to a halt. At this point the raft was disassembled or unpinned to allow the individual boxes to pass through the canal locks. The boxes would be lowered by ropes into the canal entrance. Village boys delighted in riding through the locks on the raft and helping to push open the gates of the lock.

At Bellows Falls during the busy season the rafts would often accumulate in large numbers in the bend above the canal, awaiting their turn to pass through. At times, as many as 200 rivermen could be in town. The lock tender was supposed to let the rafts through in order of arrival, but he wasn't adverse to accepting a bribe and passing a raft during the night hours so that it could reach market ahead of the others.

Rafting was hard work, but the crews also played hard at night when tied up at a village such as Bellows Falls. They would spend the evenings in ring wrestling or in challenging the locals in a game of baseball. If any musicians could be found among the crews they would have a dance; over 100 men are said to have once danced nearly all night in the square at Bellows Falls.

Once the boxes were passed through the canal locks they could be assembled again into a raft in the eddy below. But navigating the assembled raft out of the eddy could be difficult because of the swirling current below the falls. It was not unknown to have to squander a full day to work a raft free of the strong eddy currents, using extra hands to help pole against the current.

Taverns in the river villages built up a large trade with river traffic. Supplies and drink were made available, as were overnight accommodations for those who preferred such. The Thomas Durkee tavern at the mouth of the Millers River in Massachusetts was ideally situated to service river traffic. There would often be a hundred rafts backed up at the locks at Turners Falls; the immediate area would be alive with the hearty rivermen from locations all along the river.

Toll rates varied at the various canals, but they could add up considerably for the operators. Manufactured lumber would be charged in the vicinity of 25 cents to 75 cents per ton, and often double for thicker timbers. It was up to the canal operator to estimate tonnage of the load, usually calculating a ton per thousand board feet of sawn lumber. A complete raft could be estimated to weigh as much as 150–170 tons when raft logs were included, thus demanding a sizeable toll. But much of the decision regarding the amount of the toll was up to the discretion of the canal operator.

TARIFF OF TOLL,
AT
BELLOWS FALLS CANAL,
for 1818.

Every 2 Hhds Liquor, Salt or Sugar, as one Ton, at	$00,75
3 do. dry goods, grain or other light articles,	75
3 1-2 tierces Salt or every 25 bushels,	75
5 do. dry goods, grain or light articles,	75
8 Bbls. liquor, sugar, fish, salt, or provisions,	75
5 1-2 do. pot or pearl ashes,	75
10 do. Flour grain, or light Articles,	75
4 Chest Tea,	75
4 Crates Crockery Ware,	75
And for every Boat passing the Canal	2,00

ALSO.

Every 1m. ft. 1 in. boards, as one Ton,	$0,75
1-2 m. 2 in. Plank, (and so in proportion,)	75
1 1-2 m. saw'd Clapboards	75
6 m. 18 in. Shingles,	75
1-2 m. Hhd. heading as one Ton,	75
2-3 m. do. Staves,	75
1 m. Barrel do.	75
Long Sticks passing Locks, each	75
do. do. drawn from pond, do.	75
Saw logs *under* 20 ft. in length, each 20 cts. or 75 cts 1 m.	

For the accomodation of Boats passing through the Canal, not more than six boxes of Lumber will be allowed to enter the head, at one time—and all boxes, so coming in, over six in number, shall pay at the rate of $1,00 per Ton, estimated as above.

For the further accomodation of Boats, and to facilitate the passing of Rafts. no boxes shall exceed 54 ft. in length, 13 ft. in width and draw more than 2 feet of water.

All persons concerned, will govern themselves accordingly.

N. B. Payment will be expected on passing through the Locks.

HALL & GREEN, *Agents.*
Bellows Falls. January 1, 1818. 54

The tolls for lumber and other goods passing through the locks in 1818 were advertised in the Vermont Intelligence *and* Bellows Falls Advertiser. *Note that the toll for each mast log is 75 cents.*

Raft operator Steven Morse found a unique way to take advantage of this method of toll assessment. Morse was a well-known riverman who once had a large gang of rafts waiting to go through the locks at Turners Falls. He was told the rafts would accumulate a toll cost of $800. It was too much, he proclaimed, and then threatened to dismantle the rafts and run the logs loose over the dam during the next 24 hours. Fearing he might lose the toll, the agent dropped the price to $400; Morse let it be known, however, he would pay only $200. Failing to get a further concession, Morse employed an intimidation tactic against the canal agent. He gathered a crowd, gave a prolonged "Baptist sermon," and then took a large collection, which he donated to the poor. In appreciation, the local tavern keeper, a Mr. Henry, crossed out the accounts in his book against Morse and his crew. On Sunday evening the rafts went through the canal at Steven Morse's price.

Log rafting was a busy enterprise on the upper Connecticut River during the early 1800s. Records of the locks at the Bellows Falls canal show that in some years there was a passage of more rafts than there were of freight-loaded flat boats. By the midpoint of the nineteenth century, however, more and more of the forest products were being shipped on the newly built railroads along the Connecticut River.

One of the last operators to raft lumber down the upper river was the firm known as Baldwin Brothers of Bloomfield, Vermont. In 1849 they began construction of the Nulhegan Mills, a large sawmill at the mouth of the Nulhegan River. In May 1851 they rafted some of the first lumber they had sawn and sent it on down through the treacherous Fifteen Mile Falls.

The opening of the railroads along the Connecticut River (between 1849 and 1851) siphoned most of the freight traffic from the river, resulting in the eventual closing of the canals and the disappearance of the colorful log rafts. While the railroads brought an end to the commercial freight traffic on the Connecticut River, they did not replace the river as an avenue for log transport, for the greatest log drives of all were to be held over the next 60 years.

A RAFT TRIP OF 1854

C. W. Bliss, a West Fairlee, Vermont, merchant, related to historian Lyman S. Hayes his experience of working on a raft excursion in May of 1854. Bliss was only 18 at the time, hiring on as a cook for the entire gang.

There were 18 rafts traveling together, owned by different parties; the raft Bliss rode on was the property of a man named Richardson from Orford, New Hampshire. There were a total of

18 men in the gang, but since it took two men to handle the rough oars on each raft, nine rafts would be taken down the river to a spot where they could make connection with the northbound train. The crew would then return by railroad to retrieve the other nine rafts and bring them on down.

A board shanty was built on Bliss's raft that nearly covered an entire box, one end being used as a kitchen and dining room and the other for sleeping. The sleeping section had a large quantity of straw thrown loosely on the logs where the men slept fully clothed. It was large enough for the men to lay in two rows with their feet towards the center.

Although the raft had originated further north, Bliss was hired to come on board at the White River Falls at Wilder, Vermont, and depart at Holyoke. An old elevated-oven stove was secured in the shanty for cooking. Bliss cooked the same menu for every meal: brown bread, potatoes, beans, tea and coffee, sometimes supplemented by white bread when he could purchase it. As he stated, "The men were always sure to reach the raft on which was the shanty at meal time and at night."

Travel time between Wilder and Holyoke, the section Bliss worked, was 25 days. For his services as cook he received $1 per day, while the regular crewmen were paid $1.50 a day. The pilot, who commanded a pay of $3 a day, stood at the head of the lead raft and guided the gang through the areas of dangerous rocks, he being well familiar with the river. Among the crewmen was a large, powerful, 6-foot-4-inch half-breed Indian named Sam Flint who the boys wanted along when they visited the rough saloons in the river towns.

The ordeal of breaking up the rafts and passing them through the locks at the Bellows Falls canal took three days. The 18 rafts were made up of 108 boxes. Over 500 people gathered at the canal and on the riverbanks to watch the competition among the men as they were sculling their boxes in a race to reach the locks first.

Upon arrival at Holyoke the raft owners were unable to make a sale of the lumber and logs. Although Bliss departed at this point, having tired of the experience, the gang of rafts continued on down the river as far as Middletown, Connecticut.

A RAFT TRIP FROM HAVERHILL TO SPRINGFIELD IN 1850

Excerpts from Diary of Alonzo F. Niles

Alonzo F. Niles of Haverhill, New Hampshire, was just 21 years of age and had yet to see what the rest of the world looked like. He was about to begin a term as teacher at Newbury, Vermont, in the fall of 1850 but also hankered for some exciting employment for the summer. On April 27 he hired on with D. Holt to work on the Connecticut River at a wage of $14 per month. There was lumber to accumulate and rafts to be built, and once they were on the way down the Connecticut River, Niles was to be one of the three cooks for the large gang of men handling the several rafts. His family was opposed to his venture; they even held a church service to deter him, but their prayers and exhortation were all in vain. On May 31 the rafts were ready and the crew set off for Hartford, Connecticut, over 200 miles distance, with Niles aboard. The following are excerpts from the diary of Alonzo Niles (spelling and grammar left as originally written).

MAY 31—Started down river. This is my first undertaking to leave home. It is a rather hard way to travil but there are a great many things to be learned on the river, as we float slowly down the majestic stream there are constantly new objects. Passed Johnson's rocks without any difficulty. Barn's (Barron's) are in Piermont, they project into the river some distance and it is difficult getting by them. Tied up at night in Piermont, had supper about dark. The beds are composed of straw on the bottom, ticks filled with straw on top of that, and then bed quilts covered over those.

JUNE 1—Pushed off about six this morning, floated down to Orford Bridge. The hands left for Haverhill to get some more lumber, they left me here alone to take care of the shanty.

3RD—After laying by two days at Orford, floated down between Lime and Thetford, hauled up just above Hanover bridge.

4TH—Started as soon as daylight and floated down to White River Locks. I took my time to examine the locks which I never saw before. There are four pairs of gates, they all open inwardly. Afternoon, my little house has got through the locks, we did not have any cerious trouble, the stove pipe being a little too long it hit over head against the beam, compressed it. Stopped over night in the eddy below the locks.

5TH—Afternoon pushed off and went about one mile down to the other locks which are considerably larger than those above. The divisions were all tied up along the bank above the dam, five boxes were taken through, when they were going through with the last one the wall fell in on to one end of it, they will not be able to operate again for a day or two, so I have got to stay here tonight. At the head of the dam it is rather a pokerish hole to sleep in, hearing the water roling and tumbling over timbers and rocks, not more than six rods below.

6TH—Went over to Hanover village to reconnoiter. It is a very pleasant place. The settlements on the road before you get to the village are poor things and look nasty and poor. All hands have been fixing the locks today. Mr. Gordon, one of our cooks, got very badly heart today fixing the locks, a leaver fell ten feet and hit him on the thigh, he fainted, and it was fifteen minutes before they could bring him too. Felt rather worried last night for feer the rafts would get loose and go over the dam, as they sometimes do.

7TH—Took a walk to White River. After they got the locks fixed and commenced operating, they found that the sand had washed in where they run the boxes out into the stream, so they could not get them along. The men had to go into the water and plow out the sand where the water was 3 ft. deep. At half past four commenced getting supper, tosted half a bushel crusts of bread, gave it a good soaking, melted two or three pounds of butter, added what hot water was necessary, put a little salt and peper, pored it on to the tost, and pronounced it done. I then warmed up a lot of potatoes and meet, put my tea steeping, set the table and got all things ready, called the men to supper. They sent word that they should not be in until dark. Had tinboiler on the stove half full of boiling water, sit my toast on that to keep it warm. Wanted some hot water to use, so I went to take off the toast, the pan sliped and fell into the kettle, the boiling hot water flew all over one side of my face. The toast was then left soaking in the kettle. I ran to the towel and wiped my face, the skin wiped off with the water. I then ran to the soap tub and put on a good thick coat of soap, which was like putting my face into the fire. Let the soap be a while, then washed it off, put some pain extracter (which I happened to have) then put on some wet flour, left it, and went about my work same as usual. Drained the water off, made a choudder of it and the hands pronounced it good.

8TH—Got through the locks this afternoon after lying at the head of the falls three days.

9TH—Staid at the foot of White River locks last night. Started at eleven of the clock, stopt above the Quechee locks. We had rather rough river to run over today, it is very crooked and swift, there are a great many rocks and ledges projecting into the stream so that it is dangerous going round them. The men had to jump with all their might to keep the division off (there are six boxes in a division). Mr. Sanborn, a pilot, run his division on and smashed up so that it will have to be rafted over again. It is my lot to sleep at the head of dams. I am tonight within hearing of the water as it rushes furiously over the precipice, shall not sleep much for feer the raft will break away.

10TH—This morning it rained hard all night, the shanty leaked some places so the men got wet. The waves run high and are continually beating against the side of the raft, and frequently on it. The men have been taking the rafts through the locks, they take through one box at a time, and then put them into divisions after they get through. They put on two sets of men to get them out of the eddy where the water runs round and round, and if not very well managed, the raft will go round with it, and perhaps smash up against the rocks. Two divisions run together this morning and smashed the top loading off from one of the boxes, one of them got out, the other continued to whirl round until they got on sixteen men, who by laying out all their strength and exerting every nerve and musle to its utmost power got out into the stream. The men take them down a mile or two and then come back a foot on the other side of the river, so I have to take the boat and go over after them.

Unhitched from above the dam and went through the locks above five of the clock A.M. It takes nearly fifteen minutes to go through. We went through safe only as the water ran into the shanty considerable. The water is rougher here than it was at White River. The shanty rocks to and fro like a cradle when the swells come in contact with it, the water pours over the dam with great fury and then rushes down amid the large rocks with such thrashing and tumbling that the roaring may be heard to some distance.

11TH—Staid at the foot of Quechee locks last night, had fifteen to breakfast this morning. Called on Isaac Gordon this morning, my neighbor cook. He is very intelligent young man about my age, sings and plays on a violin tolerably well. His father is also cooking with us, so there are in all three families of us, sometimes we are near neighbors and sometimes twenty miles apart.

12TH—After staying at the Quechee locks three nights, we have started once more toward Old Hartford, had a hard time getting out of the eddy, the water being so high they had to truck each division out separately, which took forty men and a yoke of oxen. Got my dinner in good season because I had some ways to carry it, the rafts being more than a mile apart. Left Quechee locks in the morning at eight o'clock, passed to Bellows Falls, where we tied up at nine of the clock in the evening.

13TH—Laied at the head of the Bellows Falls all day.

14TH—The cannal and locks are on the Vt. side. It is nearly half a mile through them, a part of the way through solid ledge so rising considerable above the surface there are nine locks, sixty feet long, sixteen wide nearly. The box is got into the first one, a pair of gates shut behind, and a pair of small gates opened in the large ones before to draw the water off, so the box settles down about 8 ft then the large ones are opened, and the box goes into another, the water is drawn out of that and so on.

15TH—Went through the locks last night, we were about one and a half hours going through, during the time I got supper and eat it while they were in the opperation. Staid at the foot of the falls. This morning the hands all but two started off with a part of the lumber to be gone until Monday, so I shall have nothing to do but work for three. Went into the village today and bought fifty weight Cod-fish and to see about getting some baking done, also settled for the washing bill which was three dollars ninety cents, got four boxes of mustard and teapots fixed.

18TH (TUESDAY)—After laying at Bellows Falls six days, we have once more started towards Hartford. We pushed off as soon as light, stopt in the lower part of Hinsdale over night. Going over swift water, which is about eight miles, we had nine new hands go to pilot us over, it being dangerous and rough river to run over.

19TH—Got to Millars Locks at noon, tied up and took dinner there, soon after the lock was put in to operation, got the boxes all operated about six o'clock.

20TH—Staid at the head of Millers locks last night. The dam is sixty feet high, the highest dam on Connecticut River. There is a small Mountain in the middle of the river a little higher than the dam, against which the dam is built.

23RD (SUNDAY)—Left Montique last Friday noon, had a pleasant time until Saturday noon when the wind began to blow us on shore, where we had to wait until it ceased. It also rained hard all that afternoon and night. Arrived at Hadley cannal this noon. Opposite us is the new city, goes by the name of New Ireland.

24TH—Staid in the cannel that goes through New city last night. This cannel is three miles long and is now fitted the whole length with lumber. It is dug through solid ledge most of the way.

26TH—Two men were fishing below the dam. They had about five hundred weight of shad in the boat and were going ashore, the boat hit a rock and was upset. The boat was dashed to pieces, one man barely reached the shore, the other after laying out all his power and strength to get out was forced into the rapids and besides being drowned was badly gamed and brused up.

27TH—Staid in Cabots Ville last night, it also has the name of Chicopee City.

28TH—Staid in the same place last night. The wind blew from the south yesterday so we could not run, the hands were at liberty to go where they pleased. Some spent the day in the bowling alley gambling, and some staid in the shanty with me. Come night three or four were so drunk they could scarcely go. One of them abused a couple of Irishmen and would have got whipped himself if the rest of the hands had not helpt him. The Irishman got a search warent made out and calculated to take him that night. A little past eleaven o'clock we were routed, two men stood in the door with a lantern and inquired if the Captain of the boat was on board. Answered that he was, said one of his men had been abusing a man, and he wanted to take him. The bos got up and said if there was a man in there who had been breaking the laws of peace to come out and show himself. They all crawled out one after another, each man went to the accuser and inquired if he as the one. They were all pardoned but one, the Irishmen said he believed he was the chap, only he had shaved since, but he was mistaken, the right man was in the straw under a bedtick. The sherieff went away and the rogue soon after, for fear they would come again.

29TH—Arrived at Springfield City about eight of the clock yesterday morning.

JULY 1—Arrived at Thompsonville this morning, eight miles distant from Springfield, stopt there through the day to unload some lumber, was busily employed myself in getting provisions for the shanty, have but eight men to cook for now, it takes me only about four hours a day to do my work. Four or five of the men had a regular built train today, because the overcier happened to be gone, they got just drunk enough to be jolly. Two of them go to fighting. At first they used nothing but their fists, but soon they had resorted to clubs and everything they could get hold of. The people gathered upon the bank of the river from all the near settlements, the whole crue collected, and the sober ones endeaved to pear them and stop the fighting, but one of them stepts his bare foot upon an ax and cuts it more than half off, another had to go for the doctor, the people on the shore then came a board and stopt them. The lame man went home sober enough, rest of them busily to their work and I think will be lame tomorrow.

2ND—Left the village of Thompsonville for the cannal. Have got a new set of hands, the old ones have gone back to Springfield. Think it is a good change for I have got eight good ones now. Bought a lot of fresh beef this morning, paid ten cents a pound. A lot of bread for five cents a loaf, butter fifteen cents per pound, potatoes twenty-eight cents a bushel.

3RD—Have been alone today, hands all went a long, left me with my shanty in the cannal to wait for a gang that is behind. This cannal is six miles long, wide enough for steam boats to pass each other. There are two running from Springfield to Hartford twice a day in opposition to the Railroad. Fare is only twenty-five cents, on the cars it is seventy-five.

4TH—Independence. There is a great celebration at Springfield. Should like very well to go but cannot leave. Am now in Windsor Cannal. The steamboat went up this morning crowded full of people, with her colors flying. A message from the Overseer which requested me to prepare supper for thirteen that night, who would be there soon. Got supper ready, waited till after eleven not a man come, lay down again, did not get up until late in the morning expecting no one there. Got breakfast for myself and one other. We had Nutcakes, Sweetcake, Custard, baked pork and beans, which I prepared for Independence (not for the whole gang though). Just got my work all done up slick and was thinking which way was best to go, as I intended to look round the country some today, as I cast my eye up the cannal and saw the men all coming, hungry as bares, and wanted their breakfasts as soon as possible, had some one there to eat almost every hour in the day.

6TH—Have got in to the river again, but the wind blows so we cannot run. Had a terrible thunder shower last, the people said it was the worst they ever knew. Got tired waiting so I left the company, took the Cars and went to Hartford.

Alonzo Niles returned to Haverhill in August after being awestruck during his visits to Boston and other cities. The raft trip to Hartford had consumed 31 days. He taught school in Newbury in 1850–51 and then moved to Springfield, Massachusetts, where he became a successful grocery merchant. However, before moving from Haverhill he covertly learned to play cards in the barn loft so as to be prepared for his debut to the big city.

3

David H. Sumner

Connecticut River Lumberman

THE CONNECTICUT RIVER was not overcrowded with log drives during the first half of the nineteenth century, and certainly not when compared with what was to float down in the later 1800s. But there were a few busy entrepreneurs who utilized the river. And prominent among them was a Vermont businessman named David H. Sumner, a lumberman who used the Connecticut to transport both his raw material and his lumber products.

Moving from New Hampshire to Hartland, Vermont, in 1805 at the age of 29, Sumner purchased the Perez Gallup estate four years later and entered the lumber business. In addition to a sawmill situated at the head of a stretch of rapids known as Water Quechee Falls, Sumner acquired a canal and set of locks alongside the rapids on the Vermont side. These locks were used for passage of the commercial flat boats and timber rafts then busy in the river. The pair of locks dropped the boats just 12 feet—the smallest drop of any of the six canal lock systems built on the Connecticut River between 1795 and 1820—but they were essential to bypass the rapids. Previously, a dam was built across the river at the head end of the locks. This dam ponded the river for several miles back over some shoals, reaching all the way back to the mouth of the Quechee River.

The sawmill, situated on the ledge between the canal outlet and the river, was rebuilt by Sumner to become one of the larger mills on the Connecticut River. He also set up a large lumber sales outlet in Middletown, Connecticut, as well as other yards in Springfield, Holyoke and Hartford. It was claimed that the majority of the homes built in Hartford in the early 1800s were made of lumber sawn by Sumner.

For almost a half-century David Sumner used the Connecticut River for about all of his transportation needs. Large timber tracts were acquired in northern Vermont and northern New Hampshire along the Johns River. Sumner had a variety of partners for the various firms he created, but usually operated the lumber manufacturing under the name of D. H. and J. B. Sumner & Co. James B. Sumner was his nephew in charge of much of the sawmill activity. The Sumners later set up a second sawmill on the bank of the Connecticut River in Dalton, New Hampshire.

The sawn lumber was shipped south to Massachusetts and Connecticut, occasionally on a flat boat but more often on rafts built of logs, which were later sawn up by a mill in Connecticut. Sawn lumber, shingles, and potash would be piled as high as practical on the crude raft, and the trip south was not an easy journey for Sumner's

David H. Summner.

Saws Used in Early Sawmills

In the early 1800s the type of saws used to break down a log into lumber were the long vertical saws with an up and down motion. The saw would either operate singly or in a multiple set held within a frame. The latter were known as sash gang saws and were usually found in the large mills.

When David Sumner improved the sawmill he had purchased at Water Quechee Falls in 1805, he installed three main up-and-down saws to process the logs, plus three shingle machines. The up-and-down saws were 6½ to 7½ feet in length.

The Dalton sawmill also had two of the same type of upright saws, but in the mid-1800s, when the circular saw became developed, James B. Sumner installed some circular saws to speed up production. David Sumner opposed use of those new and different type of saws, but his nephew, James, wrote him in 1850 that his fears were groundless.

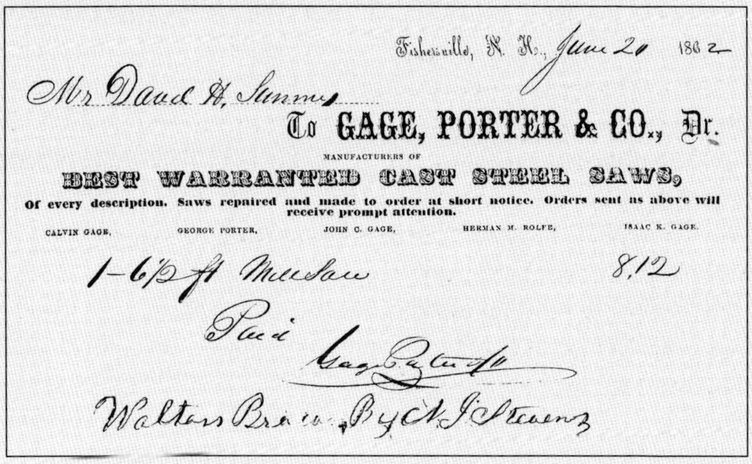

Fisherville, N. H., June 20 1862

Mr David H. Sumner

To GAGE, PORTER & CO., Dr.

MANUFACTURERS OF

BEST WARRANTED CAST STEEL SAWS,

Of every description. Saws repaired and made to order at short notice. Orders sent as above will receive prompt attention.

CALVIN GAGE, GEORGE PORTER, JOHN C. GAGE, HERMAN M. ROLFE, ISAAC K. GAGE.

1-6½ ft Mill Saw 8,12

Paid Gage Porter & Co

Walters Bro & Co, By N J Stevens

A receipt for an up-and-down saw blade, 1862.

The account ledger of David H. Sumner, 1836–1837.

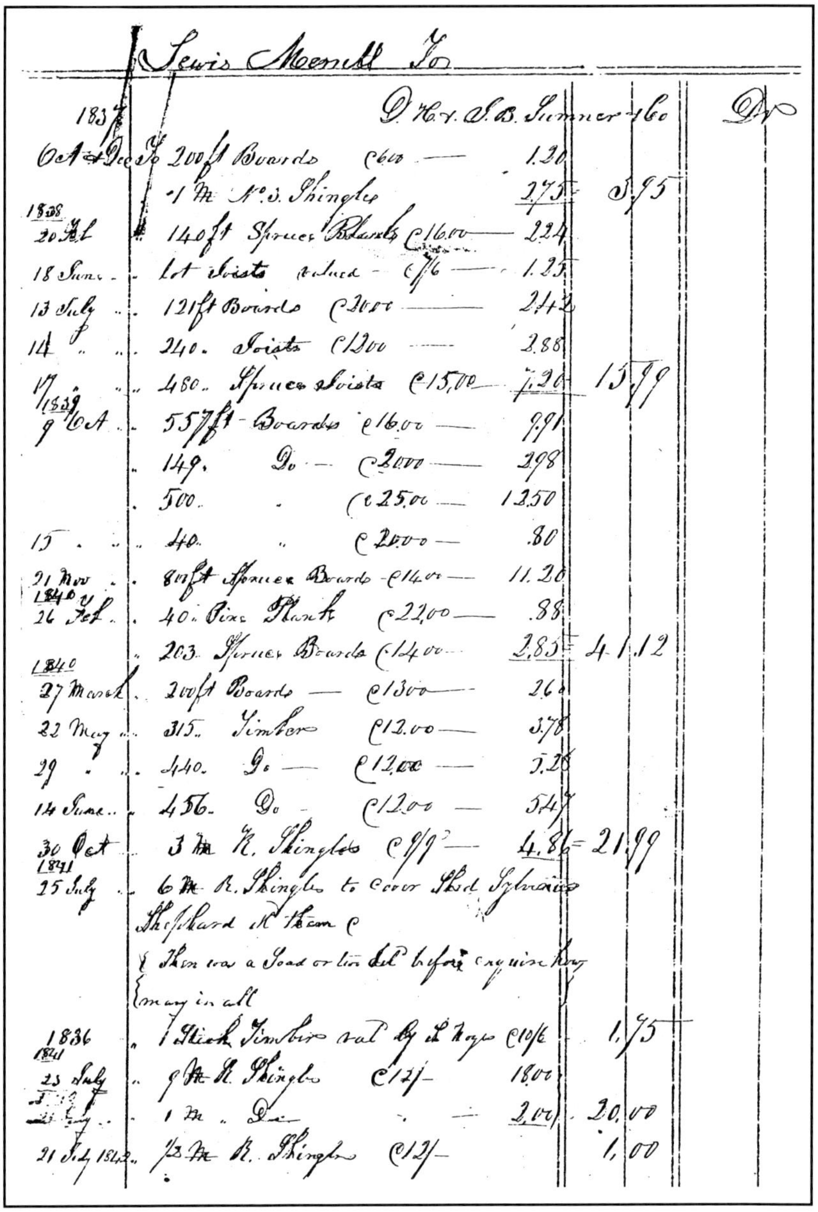

Lewis Merrill To D. H. & S. B. Sumner & Co Dr

1837 Oct & Dec	To 200 ft Boards @6.00	1.20		
	1 M No. 3 Shingles	2.75	3.95	
1838 20 Feb	140 ft Spruce Plank @16.00	2.24		
18 June	lot Joists [illegible] @/6	1.25		
13 July	121 ft Boards @20.00	2.42		
14	240 Joists @12.00	2.88		
17	480 Spruce Joists @15.00	7.20	15.99	
1839 9 Oct	557 ft Boards @16.00	8.91		
	149 Do @20.00	2.98		
	500 @25.00	12.50		
15	40 @20.00	.80		
21 Nov	800 ft Spruce Boards @14.00	11.20		
1840 26 Feb	40 Pine Plank @22.00	.88		
	203 Spruce Boards @14.00	2.85	41.12	
1840 27 March	200 ft Boards @13.00	2.60		
22 May	315 Timber @12.00	3.78		
29	440 Do @12.00	5.28		
14 June	456 Do @12.00	5.47		
30 Oct	3 M R. Shingles @1/9	4.86	21.99	
1841 25 July	6 M R. Shingles to cover Shed Sylvanus Shepard [illegible]			
	[illegible]			
1836 1841	1 Stick Timber [illegible] @10/6		1.75	
25 July	9 M R. Shingles @12/	18.00		
	1 M Do	2.00	20.00	
21 July 1842	1/2 M R. Shingles @12/		1.00	

Lumber Sawing Costs

In January 1834 David Sumner contracted with Nathaniel Shaw and William Allen to operate the sawmill at Hartland (Water Quechee Falls) for one year.

Included in the agreement was the arrangement that Shaw and Allen would pick up the logs in the booms and on the islands above and below Pulsifer's Island and Gallop Island, raft them to the mill for sawing, and deliver the sawn lumber to the boardway on the south end of the mill.

The contract price for Shaw and Allen was as follows:

For lumber sawn = 87½¢/MBF* (included cost of rafting logs to mill)

For lumber sawn at night = an additional 12½¢/MBF

For shingles made = 78¢/MBF

*MBF = thousand board feet

crew. The rafts could not be taken safely over the falls which they occasionally encountered, and to pass through the canals and locks next to the falls they had to be dissembled into the narrow sections or boxes, or in some cases completely dismantled so that the logs could be released singly over the falls, below which the raft would be assembled again. Meanwhile, the raft cargo would have to be unloaded by hand and taken around to the base of the falls by wagon.

Each box, about 60 feet by 12 feet in size, could hold as much as 9,000 board feet of pine and spruce lumber, plus a few hardwood boards such as birdseye maple, which had a small demand at the Connecticut yard. A raft with a three-man crew would carry 18 to 20 thousand feet of clapboards.

Lumber Values in the 1820s

The following values are the prices received by David H. Sumner for lumber products at his yard in Middletown, Connecticut, and other locations where sales were made.

White pine: Boards	=	\$11–\$15/MBF
Clear	=	\$35–\$40/MBF
Spruce & Hemlock: "Merchantable"	=	\$16–\$18/MBF
Boards	=	\$9–\$11/MBF
Clapboards	=	\$10.50/MBF
Shingles	=	\$3–4/MBF
Birdseye Maple Lumber	=	\$60/MBF
Chestnut and Oak	=	\$10/MBF
Framing Timbers	=	\$8-\$9/MBF
White Oak Timber—long length	=	\$15/MBF
Pine mast log	=	@\$50–\$75/log
Potash: "First Sort"	=	\$95/ton
"Second Sort"	=	\$23/ton

Pine shingles were illegal to sell unless inspected. Orders were received for pine timbers 26 feet long or even longer, which would be free of all knots and defects.

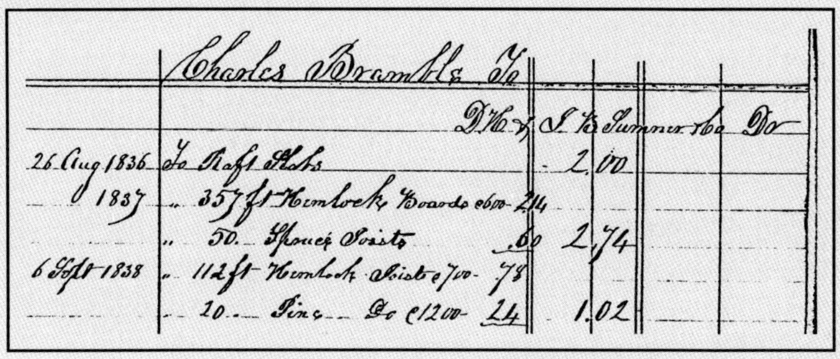
Charles Bramble To

D. H. & S. H. Sumner Co Dr

26 Aug 1836	To Raft Stakes		2.00	
1837	.. 357 ft Hemlock Boards @ 600	2.14		
	.. 50 Spruce Joists	.60	2.74	
6 Sept 1838	.. 112 ft Hemlock Joists @ 700	.78		
	.. 20 .. Pine .. Do @ 1200	.24	1.02	

A bill for lumber sold at Sumner's sawmill during the late 1830s.

LOOSE LOG DRIVES

Virtually all of the sawlog supply for Sumner's sawmills arrived from further north via the Connecticut River. White pine, spruce and a limited amount of hemlock logs were floated down to the Hartland mill for manufacture, and some logs were even floated all the way down to Connecticut to other sawmills. Then in the 1820s David Sumner acquired some timber tracts in Dalton, along the Johns River, and another sawmill was either purchased or built on the Connecticut River near the mouth of the Johns River. Logs were dumped into the Johns and collected at a boom near the mouth of the river, destined for either the Dalton or Hartland mill.

In April of 1823 Sumner's crew was running two million board

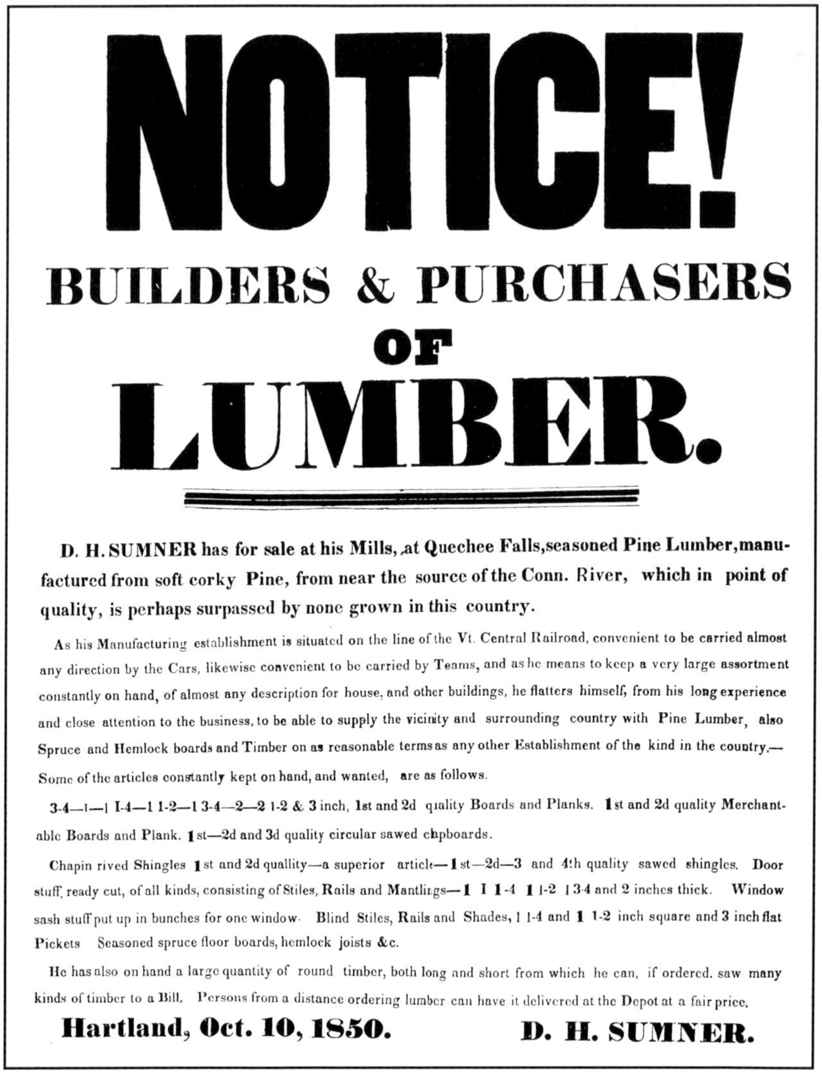

NOTICE!

BUILDERS & PURCHASERS OF LUMBER.

D. H. SUMNER has for sale at his Mills, at Quechee Falls, seasoned Pine Lumber, manufactured from soft corky Pine, from near the source of the Conn. River, which in point of quality, is perhaps surpassed by none grown in this country.

As his Manufacturing establishment is situated on the line of the Vt. Central Railroad, convenient to be carried almost any direction by the Cars, likewise convenient to be carried by Teams, and as he means to keep a very large assortment constantly on hand, of almost any description for house, and other buildings, he flatters himself, from his long experience and close attention to the business, to be able to supply the vicinity and surrounding country with Pine Lumber, also Spruce and Hemlock boards and Timber on as reasonable terms as any other Establishment of the kind in the country.—Some of the articles constantly kept on hand, and wanted, are as follows.

3-4—1—1 1-4—1 1-2—1 3-4—2—2 1-2 & 3 inch, 1st and 2d quality Boards and Planks. 1st and 2d quality Merchantable Boards and Plank. 1st—2d and 3d quality circular sawed clapboards.

Chapin rived Shingles 1st and 2d quallity—a superior article—1st—2d—3 and 4th quality sawed shingles. Door stuff, ready cut, of all kinds, consisting of Stiles, Rails and Mantlings—1 1 1-4 1 1-2 1 3-4 and 2 inches thick. Window sash stuff put up in bunches for one window. Blind Stiles, Rails and Shades, 1 1-4 and 1 1-2 inch square and 3 inch flat Pickets. Seasoned spruce floor boards, hemlock joists &c.

He has also on hand a large quantity of round timber, both long and short from which he can, if ordered, saw many kinds of timber to a Bill. Persons from a distance ordering lumber can have it delivered at the Depot at a fair price.

Hartland, Oct. 10, 1850. **D. H. SUMNER.**

An 1850 broadside advertising David Sumner's wood products.

feet of logs down the Johns River, a mixture of short and long timber. But there was a problem that year; mixed in with Sumner's logs were another million board feet belonging to five or six other owners, a sizeable log drive for that era. Since there was no drive coordination among the logging companies, a major sorting operation had to be set up for delivery to the proper locations.

Sumner's logs were branded with JB, which was chopped into the side of the log with an axe, and with a black S painted on the log end. JB stood for David's nephew, James B. Sumner, who was a member of the firm and in charge of the Dalton operation.

Sumner had a riverman that year named Chambers, from nearby Lunenburg, Vermont, who drowned under strange circumstances. He had been riding down the Johns River on a log when he fell off, but was able to eventually struggle to shore. As Chambers stood up in the shallow water, the log drive boss, William Taylor, suggested he get up on shore and rest. Instead, Chambers turned

Driving the upper Connecticut River.

around and for some unknown reason walked into deeper water and disappeared. His body was found the next day about 150 feet down river.

The following year, 1824, the company drove two million feet of short timber and another half million feet of long timber from the Johns River. The expression "long timber" referred to either mast logs or extra length logs meant to make long framing timber.

The Johns River was reported to be an easy river to drive, except when the Connecticut River had a high water flow; backwater into the mouth of the Johns River would then make if difficult to get the logs out into the Connecticut. That was not the situation in May of 1823, however; the logs were still in the Johns, waiting for the Connecticut to rise and flow enough water over the falls to carry the long timbers over.

Although white pine was not as plentiful in the North Country of the Connecticut River as it was in the lower part of the river valley, there were a few stands of pine along the Johns River. Any pine tree of suitable size and quality was cut into a mast log, for logs for ship booms, masts or bowsprits still brought a premium price. The mast logs were mixed with the sawlogs on the river drive.

Sumner customarily let out all of his woods work to be done under contract, not only the timber felling and skidding, but also the river driving. Charles Johnston and Morley Chamberlain of Newbury, Vermont, were frequent contractors, such as the job they took in 1825 to drive two million board feet of logs from the Johns River down to Quechee Falls (Hartland) for $140.

However, Sumner's representative, N. P. Page, was not satisfied with the efforts of Johnston and Chamberlain to keep the logs moving during that 1825 drive. He reported that the contrac-

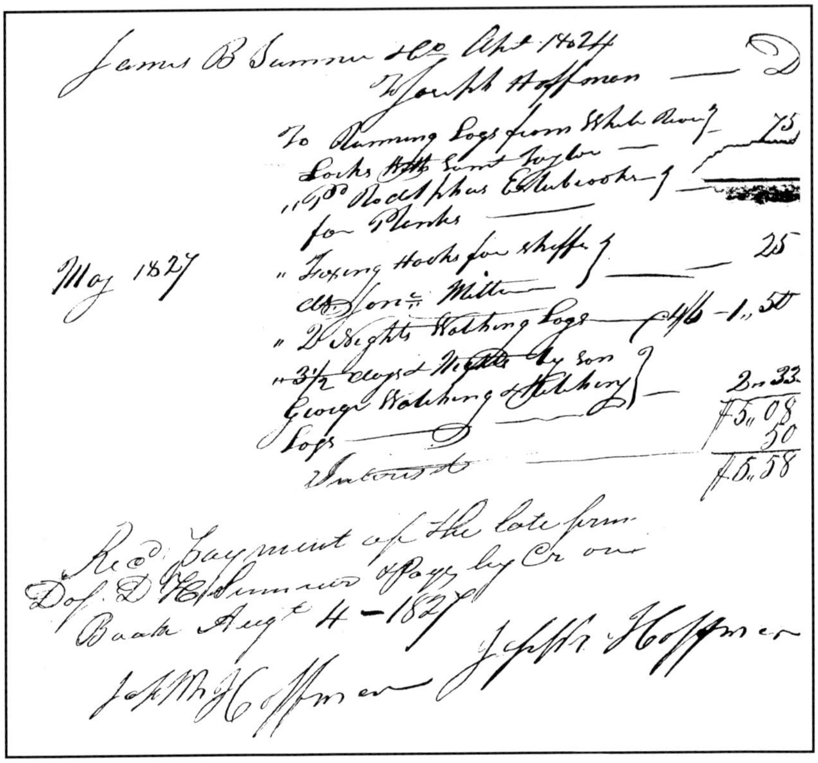
James B Sumner & Co Acct 1824 Dr
To Joseph Hoffman —
To Running Logs from White River Locks — 75
May 1827 " Fixing Hooks — 25
" 2 Nights Watching Logs — 4/6 — 1,50
" 3½ days my son George Watching Logs — 2,33
$5,08
Interest 50
$5,58
Rec'd Payment of the late firm D H Sumner Book Aug 4 — 1827
Joseph Hoffman Joseph Hoffman

Joseph Hoffman drove logs down from White River Locks in May 1827, for which he was paid $5.58 according to Sumner's receipt.

tors had a good lot of logs, including 50 long mast sticks from the Johns River, but many of the logs were becoming stranded on the islands below Barnet, Vermont. At Baird Falls, a large jam of about 200,000 board feet had to be picked apart by 15 to 20 men using rigging blocks. The chore took a full day and cost Sumner the hefty sum of $8.

A few cases of log theft were encountered also during this drive, especially along the riverbank at Haverhill and Orford, New Hampshire. A certain W. Root was discovered to have taken up to 13 logs, but was persuaded to put all back afloat except seven, which were in a "little narrow gut or hog hole" where they couldn't be retrieved. Root was charged three dollars.

After the logs passed through the stretch known as Fifteen Mile Falls, it was customary to sort out the valuable mast logs and assemble them into a raft for passage on down to the Connecticut market. Contractors Joshua Howard and George Banfield of Haverhill had charge of the 1826 log drive and were paid 38¢ per stick to sort out any mast quality logs over 40 feet in length and secure them in raft assembly.

In later years Sumner and associates acquired some additional timber holdings in Northumberland, New Hampshire, along the Upper Ammonoosuc River. In 1846 and 1847 contractor James Richey and James Abbot agreed to cut (and river-drive) the pine, spruce and hemlock logs down both the Ammonoosuc and Connecticut Rivers. Each log had to be branded with DHS, and cut with an axe on each log end and in the middle. Sumner paid them $3.75 per thousand board feet for the pine delivered to the Dalton mill and $2.75 for the spruce logs delivered to the head of Fifteen Mile Falls, en route to the Hartland mill.

SUMNER'S LOG RAFTING

In addition to feeding the sawmills at Dalton and Hartland, the Connecticut River also carried many of Sumner's sawlogs down to Massachusetts and Connecticut, where the lumber was sawn under contract for Sumner. For this long trip through a region more heavily populated and over a river course punctuated by log dams, canals and locks, the logs would usually have to be fastened into rafts. Piled on the rafts would be products of the sawmill: lumber, timbers, shingles, lath, clapboards, and occasionally potash. In 1832 a raft consisting of three "boxes" or divisions was reported to have 26,614 board feet of lumber and shingles en route from the small mills along the river.

This rafting down the lower Connecticut River was also done under contract for Sumner. Riverman Lemuel Gilson, one of the contractors, made a number of trips for which he received $4 per thousand board feet for lumber transported and about $2.70 for clapboards and lath. Payment for logs taken down was $5 per thousand board feet plus an extra $2 for each long mast stick that was too long to go through the locks at Bellows Falls and had to be coaxed over the falls.

A river trip in May of 1824 included a load of shingles piled on a raft made of mast logs of excessive length. Two days were spent unloading the shingles and carting them around to the foot of the Bellows Falls while the raft was dismantled and the mast sticks turned loose to go over the falls. The kitchen and supplies were carried on a smaller raft box that was able to pass through the locks.

Low water usually meant extra work for the crew for the logs needed a good water flow to pass over a dam or the falls. The water was down during a June 1824 trip, resulting in many logs becoming lodged in the falls at Millers Falls and at Hadley, Massachusetts. Getting the logs free was a difficult task, and where possible, teams of oxen were used to get the job done. In one instance at Millers Falls, a skiff upset in the river and one man was drowned.

Another trip down the Connecticut in the fall of 1824 featured 28 raft boxes, all loaded with wood products. When the rafts arrived at the South Hadley canal it was a Sunday morning, and the canal crew refused to work on the Sabbath; the crew on the

George H. Miller was one of the contractors who assembled log rafts and took them to Connecticut for D. H. Sumner, a trip of about 30 to 36 days as shown on the expense accounts for separate rafts during October and November of 1847. Expenses included food items, candles, lodging and meals, labor and extra hands over swift water at $1.70 per person daily.

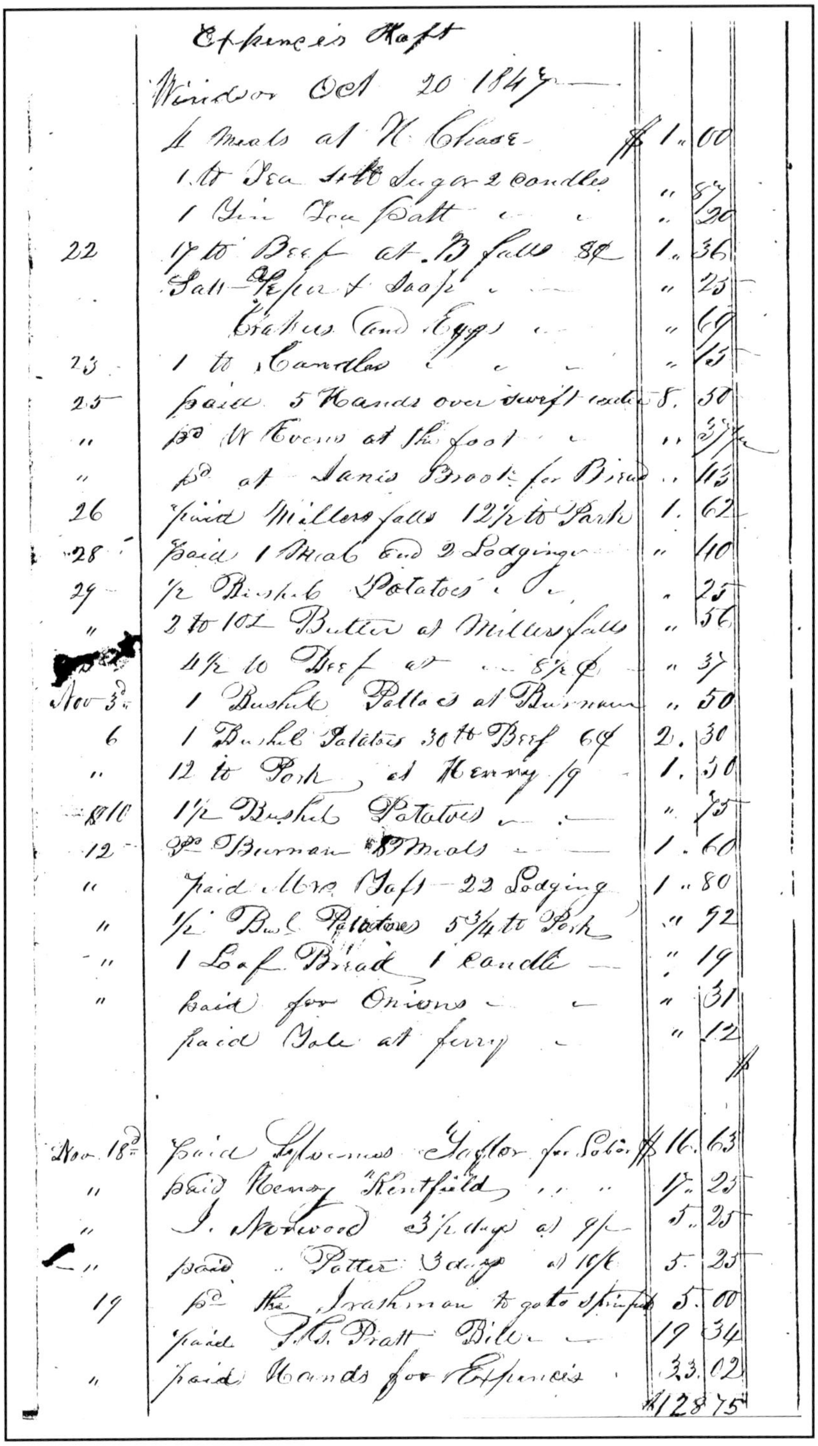

Expences Raft

Windsor Oct 20 1847

	4 Meals at H. Chase	$1.00
	1 lb Tea salt Sugar 2 candles	.87
	1 Gin Tea Salt	.20
22	17 lb Beef at B falls 8c	1.36
	Salt Pepper & Soap	.25
	Crakers and Eggs	.69
23	1 lb Candles	.15
25	paid 5 Hands over swift water	8.50
"	pd W Evans at the foot	.37½
"	pd at Janis Brook for Bread	.13
26	paid Millers falls 12½ lb Pork	1.62
28	paid 1 Meal and 2 Lodging	.40
29	½ Bushel Potatoes	.25
"	2 lb 10 oz Butter at Millers falls	.56
	4½ lb Beef at 8⅓ c	.37
Nov 3	1 Bushel Pottaes at Burnam	.50
6	1 Bushel Potatoes 30 lb Beef 6c	2.30
"	12 lb Pork at Henry 19	1.50
8/10	1½ Bushel Potatoes	.75
12	pd Burnam 8 Meals	1.60
"	paid Mrs Taft 22 Lodging	1.80
"	½ Bush Potatoes 5¾ lb Pork	.92
"	1 Loaf Bread 1 candle	.19
"	paid for Onions	.31
	paid Toll at ferry	.12
Nov 18	paid Lafayette Taylor for Labor	$16.63
"	paid Henry Kentfield " "	17.25
"	J. Norwood 3½ days at 9/-	5.25
"	paid Potter 3 days at 10/6	5.25
19	pd the Irishman to go to Springfield	5.00
	paid S. H. Pratt Bill	19.34
"	paid Hands for Expences	33.02
		$128.75

rafts wouldn't work during the evenings either. But the raft boss hired the lock tender to work extra hours and after some coaxing the raft crews took hold also; ten boxes went through the canal that Sunday evening. Then the rafts all went safely over the falls at Willimansett, one at a time.

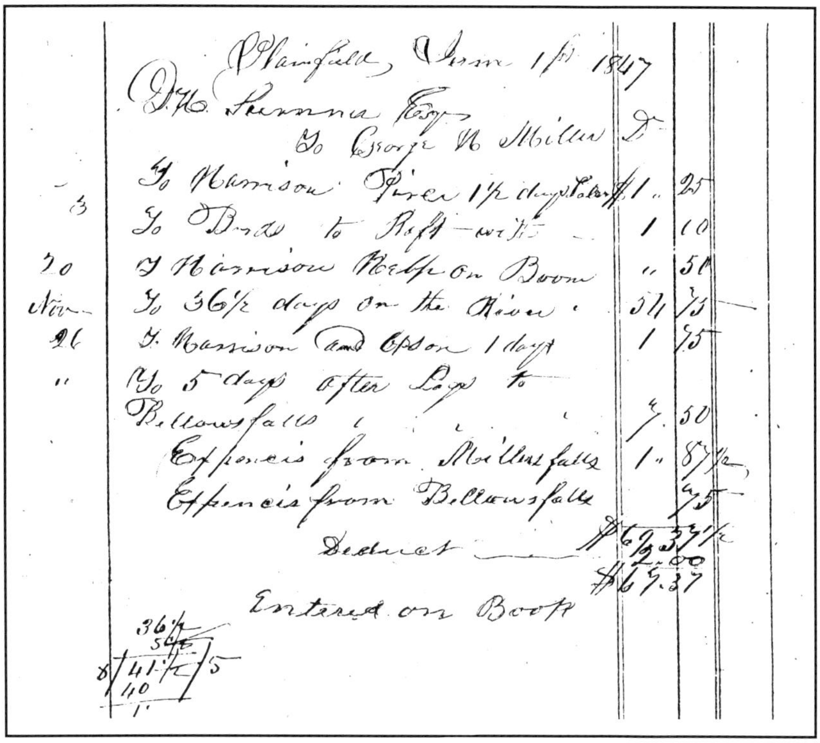

Plainfield, Jan 1st 1847
D. H. Sumner Esqr
To George W. Miller Dr
To Harrison Piper 1½ days Poles $1.. 25
To Boards to Raft – with – 1 10
20 To Harrison Help on Boom .. 50
Nov To 36½ days on the River 54 75
26 To Harrison and Upson 1 days 1 75
" To 5 days after Logs to Bellowsfalls 7 50
Expences from Millersfalls 1 87½
Expences from Bellowsfalls 75
Deduct $69 37½
2 00
$67 37
Entered on Book

PIRACY ON THE RIVER

Log thieving was a constant problem along the river for Sumner. It was especially difficult because the small crew that traveled with the rafts could not tend to all the logs when turned loose. It was not uncommon for logs to get stranded on private land when the water level dropped and then to remain there until the water rose again. The temptation to purloin the stranded logs was too great for some of the locals. This was especially true with any of the high quality mast logs.

Sumner was constantly sending a company representative down the river to search for stolen logs and to then hopefully extract payment. In 1824, agent Page collected $5 from a man in Windsor, Vermont, for a log he had taken, and within a week another $5 from a man in Westminster, Vermont, who had cut the log into bolts for shingles manufacture. Page said the fee was "trifling but the man was poor with six small children and was very humble."

Company man Albion Taylor found that Sumner's logs had been detained and stolen in many places by the river pirates. The thieves would cut out the company marks and then loudly proclaim that they had no fear of David Sumner. Even the ferry operator above the Windsor Bridge was known to have pulled in a few logs.

This thieving went on for many years. In 1840 John Kimball was

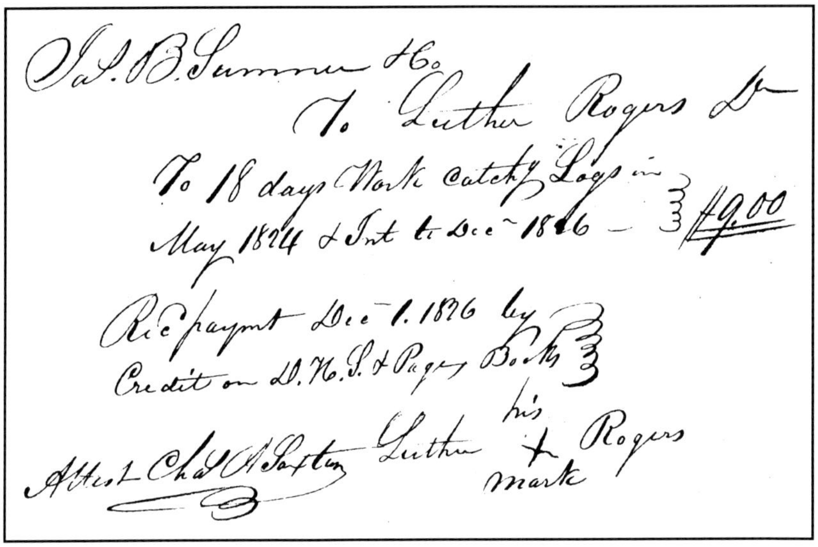

Jas. B. Sumner & Co.
To Luther Rogers Dr
To 18 days Work catchg Logs in
May 1824 & Int to Dec 1826 — $9.00
Recd payment Dec 1. 1826 by
Credit on D. H. S. & Pages Books
Attest Chas A Sexton Luther his + mark Rogers

Luther Rogers spent 18 days catching wayward logs for Sumner, which brought him a wage of $9.00. Note Rogers' signature mark on the receipt.

once searching in the Putney area for logs stolen by the river pirates and at Lord's Ferry found at least one that had Sumner's mark on it. However, he reported that most of the timber must have floated through the area at night when the pirates were too drunk to catch any of it.

The thieves had their special hiding places where they could pull a log and quickly cut it up into shingle bolts. Though the hideouts were not always easy to locate, Page once discovered a secreted pull-out and charged the thieves $30 for two stolen logs found with the company mark (IB) still on the end. That was in 1824. Seventeen years later (1841) the pirates were still at their nefarious work. That year, Kimball was able to locate two top quality mast logs that had been pulled 1,200 feet from the river bank. Both of the pine logs were 24 inches in diameter at the small end, one 75 feet long and the other 24 feet, and the company marks had been hewn out of the valuable logs. In an attempt to catch the unknown scoundrels who'd taken the logs, Kimball hid himself nearby, hoping to catch the thieves when they returned to retrieve the stolen timber. Whether Kimball was successful in tracking down the log snatchers remains unknown, unfortunately.

Local sawmills also played a part in the ongoing timber piracy. Albert Hubbard was in charge of an 1825 river trip when he discovered 30 to 40 white pine logs pulled up on Wells Island, below Bellows Falls. Included among the logs were several fine mast sticks all still bearing the IB mark. Hubbard determined that the thieves had cut open a boom that a previous crew had set up at Wells Island. The three thieves were soon thereafter discovered, along with more stolen company logs found over at Snow's sawmill in Chesterfield, New Hampshire. The axe marks had been chopped out, but the paint still showed on the log ends.

The thieves had given particular attention to the valuable mast logs. They had to leave one particularly large log at the river because they couldn't roll it over; the log was four-and-a-half-feet in diameter. At the Chesterfield mill, one mast log, which had been 80 to 90 feet long, was cut into three logs. The high quality wide boards sawn from another log still showed the company paint marks on the ends. The value of the short logs taken to the mill was $150 and each long mast log was valued at $75. The case eventually made it into the courts.

Despite the threat of lawsuits, however, some sawmill operators along the river continued to cooperate with pirates in a few cases, and Sumner was never free of the problem. Even in 1841 there were reports of thieves taking up Sumner's timber from the river between Bellows Falls and Walpole and drawing the logs to Uriah Newton's sawmill in Walpole village.

David Sumner continued to raft lumber products down the Connecticut River until 1849, the year the railroad was built along the Connecticut River valley and began to haul his lumber and shingles. Around 1857, the sawmill at Hartland's Water Quechee Falls (later known as Sumner's Falls) was carried off by a freshet

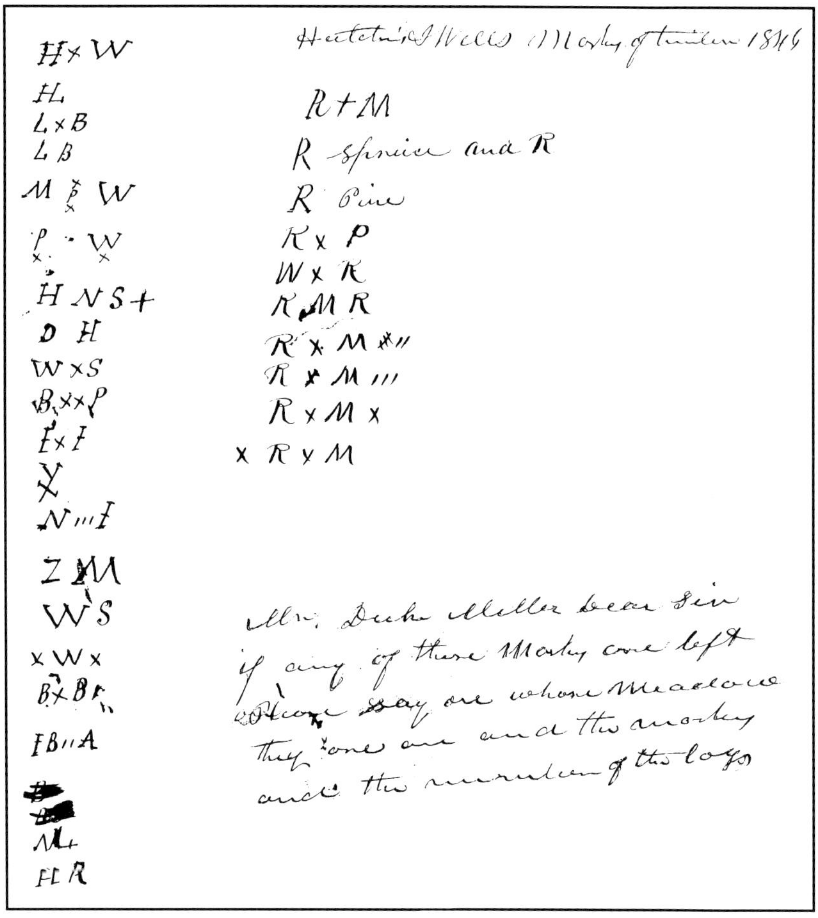

Dick Miller was sent down the river in 1846 to search for logs that had been stranded or stolen. He was to search for logs with these brand marks and report back the locations of the meadows where found.

(high water), along with the dam. Sawmilling activities by the company continued for several more years (possibly until 1865). Sumner, meanwhile, died in 1867 at the age of 91.

Nearly two decades after Sumner's death, there was a gathering of 3,000 people at Sumner's Falls to celebrate what was proclaimed to be a great future for the former site of Sumner's sawmill. On that day, September 29, 1885, paper manufacturers D. H. and J. C. Newton of Holyoke, Massachusetts, announced their intentions plans to build a hydroelectric plant and a pulp mill at the falls. The Newtons had just been victorious with an 1882 legal battle against the Ottauquechee Woolen Co. and now had approval to construct a dam 400 feet long and 25 feet high that would generate over 5,000 horsepower. Nothing ever resulted from any of Newton's rhetoric, however.

At things turned out, Sumner Falls was never again tamed and it remained one of the most dangerous locations on the Connecticut River for the large log drives of the later 1800s, and the site of many future drownings of rivermen.

PART II

1860–1949

4

The Major Log Drives Begin

1869–1878

IN THE YEARS 1845 through 1852 the mode of transportation along the Connecticut Valley was radically changed, forever. The railroads were built up the valley. Rails were laid up along the Connecticut River, connecting lower New England with northern New York and Canada via Vermont.

Shipments of lumber and other merchandise ceased to travel by flat boat and raft as industry quickly adapted to moving freight by rail. The canals and toll locks on the river suddenly became obsolete for any further commercial traffic on the river.

There was one item of traffic that was not affected by the railroads, however—the log drives. In fact, the river drives would increase in quantity in the ensuing years. The railroads were not able to replace the river transport of logs because the rails didn't reach into the remote timber country; the streams and rivers did. It would be another three or four decades yet, but the Connecticut River was to eventually host what might be the greatest log drives the country has ever seen.

Smaller loose log drives had continued along the Connecticut during the mid-century, usually short length drives headed for a sawmill located along the upper Connecticut River. Then in the late 1860s efforts were begun to float logs all the way from the river headwaters to locations in Massachusetts or Connecticut. It was the beginning of major efforts yet ahead.

The drama began one spring day in 1869, soon after the ice had gone out of the up-country lakes and the river was free to flow openly, swollen by melting snow. O. F. Richardson of Clinton,

The upper end of the Connecticut River has a high water flow as soon as the ice breaks free.

High water in the spring doesn't always cover over the boulders and gravel bars, as evidenced here as a riverman works to free up a jam.

Maine, had taken a contract with some Massachusetts lumbermen to float 1.7 million board feet of logs down to Holyoke, Massachusetts, and to then use some of the timber to construct a new dam across the Connecticut River. Richardson looked to some seasoned rivermen of Maine, the "Bangor Tigers" of Penobscot River fame, to handle the task, and he brought 36 seasoned veterans over to New Hampshire with him, including one man who was 59 years old. The crew built a flat boat with side boards and a canvas covering to convey their supplies down the river with them.

Unfamiliarity with the Connecticut River resulted in some problems for Richardson's crew. They would sometimes mistakenly allow the logs to collect in the various eddies below the dams or rapids and would have to work two shifts through the night to laboriously draw the logs back into the current.

Richardson's log drive became quite a novel spectacle for the inhabitants of Bellows Falls for decades had passed since they had witnessed logs of any quantity coaxed over the Bellows Falls dam and off of the ledges down below. Furthermore, the locals had never seen or even heard of such a daring process as that used by the Maine rivermen to break logjams. The men would form a single file into the river, holding hands for steadiness and security, and the man at the far end, the "post of honor," had the dangerous job of pushing the individual logs back into the current. (During the 1869 drive, one man was reportedly swept into the rapids but was able to gain control, even tossing his axe up onto the logs to save it.)

The rivermen from Maine showed that large volumes of logs could be moved down the Connecticut River, over the dams and ledges, to a destination far below. Richardson delivered his logs within the 100 days allowed on his contract and built the new timber dam with time to spare.

There is also an historical reference to a log drive on the river in 1868 when a contractor named Eli Hanson is said to have made a timber cut on the lower end of Indian Stream in far northern New

The log drive just above Wilder, Vermont.

Hampshire and driven the logs down the river to Massachusetts. The *Bellows Falls Times*, however, in commenting about Richardson's 1869 log drive, stated that Richardson's was the first log drive to pass through since the era of log rafting many years previous.

A year later, in 1870, Richardson was back again, this time rebuilding the large, old 1787 dam at Bellows Falls, which consumed about two million board feet of logs driven down from Wells River.

By 1872 the down-country saw millers realized the potential and feasibility of relying on spruce and pine floated down from the North Country, thus log drives began on a large scale. Fifteen million board feet of logs came down the Connecticut that year to a sawmill at Holyoke, poked along by a crew of 75 rivermen, for the most part still recruited from the State of Maine.

Next, the down-country sawmills began to work together on a joint effort at log driving. In 1873 three different Holyoke lumber concerns joined forces to push down a log drive of 26 million board feet with a crew of 150 men who were now getting to know the peculiarities and hazards of the Connecticut River. Harvey and Win Schoppe, who were later to gain fame on the river, contracted for a large timber cut that winter near South Bay on First Connecticut Lake.

The first large log drive fatality of record occurred in 1874, although there probably were some previous drownings. In 1874 a riverman drowned at a spot known as Rhoade's Falls (location unknown), and two other men drowned near White River Junction, Vermont, when a boat with a crew of six capsized in the current while attempting to cross from one log jam to another. River

Problems arise at the Ledyard Bridge near Hanover, New Hampshire.

Large crews were required along the upper portions of the Connecticut River.

driving was a hazardous occupation, and many more deaths were yet to come in the surging spring waters of the Connecticut.

The 1874 drive started with 200 men at the headwaters, but had been pared down to 150 by the time the main body of the drive reached Bellows Falls. As many as 50 men were kept busy for some time at the Windsor (Vermont) bridge, breaking up a large jam and back-up of logs.

Large eddies below the dams usually created a troublesome location for the log drivers, as they formerly had for the rafts and flat boats. Logs that would gather in the calmer back-current stubbornly resisted efforts to coax them back into the current again. On the 1874 drive the rivermen attempted and accomplished a new feat for cleaning out the large eddy below Bellows Falls. A boom, formed by a string of logs chained together end to end, was built along the northerly and westerly inside edges of the eddy. Seven boats were positioned on the other side of the river, attached to a draw chain and manned by six sturdy rowers in each boat working in unison. With the aid of men perched on the boom logs, who were pushing the logs along with long pike poles, the boom slowly swept the mass of thousands of logs out of the large eddy, much to the delight of a large crowd of spectators who cheered them on.

THE YEAR OF THE ROSS DRIVE

The year 1876 was an outstanding one in the annals of log driving on the Connecticut. It witnessed the most famous of the log drives by the "Bangor Tigers" from Maine, and also resulted in a working

Logs move slowly below the covered bridge at Windsor, Vermont.

partnership that was to soon evolve into a corporation that would dominate all activity on the Connecticut River for four decades to come.

Lumberman W. S. Stevens of Hartford, Connecticut, in partnership with John Ross of the firm Ross and Leavitt of Bangor, Maine, had purchased 111,000 acres of timberland at the headwaters of the Connecticut. That winter Ross sent up 50 loggers, plus 30 teams of horses, from Bangor, added to by whatever local men were available to harvest the timber. By the end of the winter of 1875–76, Ross's crews had six million board feet of logs piled on the river bank, and a number of other companies likewise had their winter harvest ready to float—a total of 20 million board feet awaiting the spring thaw. In river lingo, 1876 became known as "the Ross year."

John Ross personally bossed the drive with 150 experienced river drivers that he'd brought down from Bangor. They were skilled men with a fierce pride in their ability and reputation. Included in this mixture of Canucks, Province men, Irish, and Yankees were 18 Indians from the Penobscot Reservation at Old Town, Maine, renowned for their ability with bateaux, the boats used by rivermen.

The first group of 70 men arrived in early April and soon realized it was going to be a tough drive. The snow melt was below normal and the river flow unusually low. The drive was taken down in separate sections, because the drive included, in addition to the Stevens and Ross logs, a cut of five million board feet for the sawmill in Holyoke, five million more destined for a mill at Bellows Falls, and a few other smaller ownerships.

The low water flow in the river created constant problems with ledges and exposed gravel bars, but the first drive reached White

River Junction on June 2, and the second major drive arrived two months later. The exposed ledges below Bellows Falls were especially troublesome, even though some of the advance rivermen had taken advantage of the low water to blast away a few of the rocks. Hundreds of spectators gathered on the river bank to watch the 65 to 70 rivermen struggle with the stranded logs. A pair of mules was used to pull the logs off of the ledges with the aid of a long cable.

On the tenth of August some of Ross's rivermen performed an extraordinary feat at Bellows Falls, one that had never before been accomplished; they rode their boats down through the treacherous rapids below the dam. Five boats passed the entire length with two men in each, one handling an oar in the prow and the other in the stern. Three of the boats dipped some water but the other two stayed dry. Such a feat had never been done before, but was made possible on this occasion by the low water and the experience of the daring men. These men had been on the river for at least 100 days by this time and had lost only one man, who had drowned at the beginning of the drive. Their courageous antics created a big sensation in many points along the river.

The first part of the drive arrived at the Hartford location about the beginning of September, where it was collected into booms. At this point the logs were fastened into rafts and towed by a tugboat to Little River, Connecticut, where Stevens, Ross and Leavitt had rented a sawmill to process the logs. The *Hartford Palladium* trumpeted the hopes of the city's lumber industry when it stated that "Hartford will no longer be obliged to get its lumber by vessels or cars from Maine and elsewhere, for it will be floated down the river to the city." Thus were the small beginnings of what would become a river monopoly.

The men from Maine were back again the next year (1877) pushing down two separate drives again totaling 20 million board feet. And again they dazzled the local folks with their feats of daring. A jam on the ledges at Bellows Falls was broken by a man let down by rope between the railroad and toll bridge to pick at the key logs and then roughly yanked up as the jam broke under his feet.

The hopes of the Hartford lumber manufacturers were apparently now being realized, because the *Bellows Falls Times* of June 1, 1877, reported on a meeting between the railroad freight agents and retail lumber dealers in Springfield, Massachusetts. The retailers were asking for a freight tariff on incoming sawn lumber that could make them competitive with the Hartford manufacturers conducting the log drives.

The 1877 drive went well compared to that of the previous year; in fact, there was too much water for the best conditions. High water caused the log boom above the Sullivan (New Hampshire) railroad bridge to break free on July 2, releasing the entire second drive of 11 million board feet to pass quickly over Bellows Falls. Many of the logs were to have been pulled out at Bellows Falls for use by the Fall Mountain Paper Company, but there was no stopping them once they broke free in the fast water.

The Connecticut River was kinder to the rivermen in 1878. The winter cut of 30 million board feet (40 million according to a report in the *Vermont Watchman* paper) had been logged and stacked streamside by one thousand loggers who had been brought into the woods by Stevens, Ross and Leavitt to harvest some spruce along Indian and Perry Streams. Starting off in early April, 250 rivermen went down with what was the largest drive to date. Captain Isaac Bowman of Lancaster, New Hampshire, took the contract to conduct the drive for $80,000. He paid his log drivers $1.25 to $2.00 per day, plus food.

The drive was a relatively easy one, with the river running at a high level. Ten million board feet of logs were sorted out and left at the large mill at McIndoe Falls, Vermont, another 10 million dropped off at a second large sawmill owned by the McIndoe Falls Lumber Co. at Mount Tom, Massachusetts, and the remaining 20 million floated on down to the sawmill the company was operating below Hartford.

There were no big problems, that is, until the crew reached the lower end of the Connecticut River where the situation took an unpleasant turn. Problems with people enjoying the river for summertime boating escalated worse than ever around heavily populated Springfield and Hartford, and there was still some commercial steamboat traffic on the lower river. Because of this major problem, Stevens, Ross, and Leavitt reluctantly came to the decision not to float any more logs through those congested areas of conflicting river uses. In the process of adjustment, at this time a new company that was to monopolize, even tyrannize, future log driving activity on the Connecticut River, was also formed.

5

The Connecticut River Lumber Company

1879–1902

RECREATIONAL BOATING was a large pastime on the Connecticut River in the nineteenth century; log driving was definitely not compatible with boating. Pressure from the urban areas along the river eventually became a discouraging factor for the Hartford lumbermen who had been using the river to provide their yearly supply of logs. As a result, the company sawmill and other assets were sold to a new group of investors.

The Connecticut River Lumber Co. (CRL Co.) was chartered in Connecticut in 1879 by the new investors and was actually organized as a successor to several older lumber companies. New York banker George S. Scott is reputed to have unexpectedly fallen into the lumber business when acquiring a large timber property around the Connecticut Lakes during settlement of a sizable debt. Scott and two other bankers, Thomas Pearsall and a man named Thorne, were the company principals, with Scott serving as the first president.

Asa Smith of Hartford was installed as the business manager and continued as such for the first four years of the company's existence. Smith was a pioneer in the lumber industry and was remembered as a kind man with whom to deal.

With the decision previously made that no further attempt would be made to float logs down past Springfield and Hartford, the company made arrangements for the use of two other sawmills along the river. The sawmill belonging to the Holyoke Lumber Co.

of Holyoke, Massachusetts, was leased, and the sawmill at Mount Tom (in nearby Northampton) was purchased from the McIndoe Falls Lumber Co. of Vermont in 1879. The Mount Tom mill had been built in 1873 by the group of Vermont lumbermen who also owned the large mill further upriver at McIndoe Falls.

The first log drive under the new management turned out to be a costly one. The drive of 1879 got hung up at Fifteen Mile Falls and it cost the CRL Co. $40,000 to get the logs through. In October of that year the company started blasting large boulders in the channel and cleared about ten miles through that troublesome stretch. About that time the CRL Co. also obtained a New Hampshire charter for the Upper Coos River and Lake Improvement Company, and with the charter as a vehicle, spent large sums clearing obstacles from streams and building dams at lakes to facilitate the log driving.

The log drives for the next four years averaged 50 to 60 million board feet per year and often employed as many as 700 men at the beginning of the drive. The company was learning how to handle the problem areas on the Connecticut River, but the time length of the drive still varied 20 to 100 days or more before the logs arrived at Mount Tom or Holyoke and were gathered into a boom. The length of the drive still depended greatly on the volume of water flow in the river. Among the reasons for avoiding delays on the river was the cost to the company, which exceeded $3,000 per day while the drive was active.

Deaths among lumbermen were common, but the 1882 season was especially deadly with six men killed on woods operations and then another five drownings on the log drive. The fifth drowning fatality that spring was a young man named Benjamin Chapman, who fell victim to the particularly high, fast water at Fifteen Mile Falls and whose body was recovered just above the Barnet, Vermont, bridge.

The 1882 drive was a difficult one because of the exceptionally high water flow, but the drive also proved to be one of the quickest ever. The flowing mass of 65 million board feet of logs reached Bellows Falls in 65 days. There, a portion of the logs was sorted out for sale to the Fall Mountain Paper Co., and enclosed there in a boom. But the high water brought trouble when the surging river waters broke the boom, releasing 200,000 board feet of logs down the river to destinations unknown.

By the 1880s the Connecticut River log drive was no longer dominated by the rivermen from Maine. By then crews were made up of log drivers from across northern New England, Quebec, and the Maritime Provinces. However, the tough rivermen from the Penobscot area continued to sign on every spring. In 1880, approximately 250 men came over from Bangor, about 50 came in 1883, and even in 1888 they numbered close to 250 once again.

The Connecticut was a more difficult river to drive than was the Penobscot River in Maine, but the pay for the rivermen was better. Wages on Maine's Machias River, for example, were as low as $1.25 to $2.25 per day at the time, whereas a riverman could earn $3 to $5 per day on the Connecticut and Androscoggin Rivers.

The Connecticut River Lumber Co. did not have the river all to itself during this era. The Turners Falls Lumber Co. of Massachusetts, one of the oldest sawmills on the Connecticut River, elected to conduct its own log drive in 1884, and the owner of the large sawmill at McIndoe Falls was floating his logs independently, also. This presented some problems for the CRL Co; the season was much too short to have to wait and share a limited space of time on the river.

When CRL president George Scott was told of the difficulties they were having with the owner of the McIndoe Falls Lumber Co., he reportedly told his North Country representatives to "go back and kill him off," according to historian Robert Pike. Upon receiving the report that the company men could not eliminate the owner's forceful presence and stubborn tactics, Scott stated, "Then he's the man we want."

In 1884, after CRL general manager Asa Smith announced his plans to resign due to failing eyesight, the company promptly hired on this forceful competitor as their new general manager. The name of this domineering man of indomitable energy and perseverance was none other than the renowned George Van

The CRL Co. Lake House at the foot of First Connecticut Lake was formerly a three-story hotel operated by Asa Smith.

Dyke. This move became Van Dyke's next major step toward what was to be his eventual domination of lumbering activity on the Connecticut River.

Upon his retirement, Asa Smith took over management of the hotel he owned on the Connecticut Lakes, a three-story structure that could accommodate 50 people. First Connecticut Lake was a sportsman's paradise at this time, yielding trout and lunge as large as 20 pounds. Salmon were also abundant at neighboring Second Lake. Thomas Chester had a camp for sportsman on Second Lake, charging $1.50 per day. The area was serviced by a daily stage from Parson's House in Colebrook, at a charge of $4.00 round trip.

In addition to the masterful task of managing the logging activities and log drives of the CRL Co., Van Dyke continued to operate his McIndoe Falls sawmill and another one further up the river at Guildhall, Vermont. In 1885 he leased a sawmill in Zealand, New Hampshire, belonging to a contemporary who was probably the most forceful timber baron to ever walk the woods of New Hampshire—James E. Henry. Van Dyke's general manager, John H. Locke, operated the Zealand sawmill.

The principal office of the CRL Co. eventually became established at First Connecticut Lake, along with a company hotel, several barns, a store, a small sawmill, and a small steamboat on the lake, the "Hartford." The winter months were busy with 400 or more men and 200 horses in the woods cutting 40 million board feet of spruce for the spring drive. Seven to eight hundred men would gather there in early spring hoping to sign on for the annual log drive. About 1890, the company office was moved down to West Stewartstown, New Hampshire, because of the railroad connections at that point.

George Van Dyke was apparently not an individual content sim-

The company store and office at First Connecticut Lake, plus some company barns, as they appeared in 1914. The store had a dance floor on the second level.

ply managing the assets of another owner. His partners in the McIndoe Falls Lumber Co. venture, for example, soon evaporated from the scene.

The later location of the company office in West Stewartstown, New Hampshire.

Possibly the New York bankers felt that they were too far removed from the area of activity; or possibly Van Dyke issued an ultimatum after two years of overall management. Whatever the motivation, Van Dyke proposed combining the assets of CRL Co. with his ownerships in order to expand the company activities. Thus it was that in 1886 the CRL Co. became the operating company for four sawmills along the Connecticut River: at Holyoke and Mount Tom, and at McIndoe Falls and Guildhall.

Van Dyke's timberland holdings of about 80,000 acres were combined with the large land ownership of the CRL Co. to give the company a substantial timber base in the North Country. Land ownership in the town of Pittsburg, New Hampshire, alone was now 125,000 acres. By the end of the nineteenth century the company was to accumulate 320 to 325 thousand acres of virgin timberland.

The assets contributed by George Van Dyke during the 1886 reorganization were enough to give him a commanding voice, and he soon became the new company president. He now reigned in an authoritative position to supervise lumbering and log drive activities on the Connecticut River, although George Scott and Thomas W. Pearsall still retained controlling interest in the company.

In 1891 the Connecticut River Lumber Co. formed a separate corporation, the Connecticut River Manufacturing Co. This new company took over operation of the sawmills at Mount Tom and McIndoe Falls, and also purchased the harvesting rights and own-

Connecticut Valley Lumber Co.
Manufacturers of
Spruce Lumber.
Mills at
Holyoke, Mass.
Mt. Tom, Mass.
McIndoe Falls, Vt.
Boston Office,
89 State St.
McIndoe Falls, Vt. Dec.14,1903.

The letterhead of the CVL Co.

ership of the logs cut from the forestlands of the CRL Co. The Connecticut River Manufacturing Co. was short-lived, however, as it dissolved about 1898 during a failed attempt to avoid legal responsibility for the washout of a railroad bridge at Windsor, Vermont.

An interesting development involving the CRL Co. also surfaced in 1895 when William A. Russell, president of the Fall Mountain Paper Co. at Bellows Falls, made it known he was interested in purchasing the extensive CRL timberlands in the North Country. The CRL Co. was interested in his offer and gave him an option to purchase the timberland, but the inclusion of Fall Mountain Paper Co. in the 1897 merger that formed International Paper Co. forestalled any immediate action on land deal.

Also in 1897, George Van Dyke made a proposal to purchase the controlling interest of CRL Co. from George Scott and Thomas Pearsall, and in December he acquired their stock for $1.8 million. As majority owner, Van Dyke then promoted a new financial venture, the sale of both the Connecticut River Lumber Co. and Connecticut River Manufacturing Co. to the newly formed International Paper Co. So intent was Van Dyke to consummate the deal that he traveled to Florida in March 1898 to discuss the deal with an ailing William Russell, International Paper's first president. Van Dyke offered the property for $2 million, but Russell and his company backed off for some unknown reason. One wonders what course history might have taken if International Paper Co. had followed through with their interest in purchasing the vast timberland ownership of George Van Dyke.

Van Dyke's influence and financial control continued to grow, and in December of 1902 the company was again reorganized, this time emerging as the Connecticut Valley Lumber Company. The CVL Company was to become a name recognized by local residents and woodsmen for decades to come.

In 1908 the Boston firm of Hornblower and Weeks floated a $3 million stock issue to strengthen the company. George Van Dyke purchased two million dollars of the stock and never relinquished control of the CVL before his death in 1909.

6

A Chronicle of the Connecticut River Timber Baron

THE ANNALS OF New England lumbering seldom fail to mention the name of the man of limitless energy who outranked all of the other old-time timber barons in aggressiveness and accomplishments, the renowned George Van Dyke. His monumental exploits date from an era when large acreages of virgin timber were frequently available for the keen and astute individual who had the drive to rise above his contemporaries. And the accomplishments of George Van Dyke have placed him head and broad shoulders above the select rank of timber barons of any era.

Born February 21, 1846, in a log cabin in Stanbridge, Quebec, to Dutch and Yankee ancestry, the fifth of eight children, he rose in 40 years to the point where his word went almost undisputed along the Connecticut River and throughout the North Woods.

Van Dyke once stated in some courtroom testimony that he had begun working in the woods at age 14, working at a lumber camp in Maine and earning wages as a chopper and log driver on the Androscoggin River. At age 19 he was in charge of a crew on the river. Ten years of "hard knocks" education gave him the experience that, together with his strong will and robust health, was to earn him the full respect of the rivermen in the later years. He never hesitated to step out onto the logs himself or seldom asked a man to carry out a task that he himself would not attempt.

Van Dyke was not a tall man, but he was stout and robust (said to have a 19½-inch collar) with a powerful voice to match. In later years, while personally bossing his large log drives, his bellowing voice could be heard from bank to bank along the Connecticut River.

In 1870 Van Dyke went to work for himself, at age 24. He made a small timber cut on Indian and Hall Streams and took the logs down the Connecticut River. The drive got hung up on Fifteen Mile Falls and at that point Van Dyke sold the logs to another lumberman.

Van Dyke knew from an early age that wielding an axe or rolling logs with a peavey was not what he would be content with for a livelihood. His keen mind was able to quickly spot a profitable business proposition or a timber purchase that he could profit from many times over. His business abilities and timber smarts soon led him on the path toward his dominating role.

At age 26 he entered into the sawmill business, acquiring, along with partners Henry and Lewis Bowman, a sawmill on the Connecticut River in Guildhall, Vermont.

In 1879 Van Dyke found himself unexpectedly in temporary possession of another sawmill on the river. He had made a $100 purchase of a timber lot in Hereford, Quebec, and parlayed that into a cut of about a million feet of logs. (Translation: The adjoining landowners probably had some of their timber removed also). The timber was put into the Connecticut River and driven down to South Lancaster for purchase by mill man John H. Locke, who had a sawmill there in the section known as Hillsborough. The logs had no sooner arrived than Locke was forced into bankruptcy, leaving Van Dyke with a million feet of logs boomed in the river and a crew of men unpaid for their winter's labor.

Undaunted by the threat of a huge loss and the fact that Locke's sawmill crew refused to work, Van Dyke arranged for the bank to take ownership of the mill and then selected a crew from among his own river drivers to operate the sawmill machinery and saw out his winter's cut of logs. The report is that the lumber market went up $3 a thousand feet that spring and that Van Dyke made a handsome profit of $10,000 on his venture. He was the type of entrepreneur who fell into situations like that.

Van Dyke continued to accumulate capital and increasingly became a major player in North Country lumbering. Having tasted the profits of sawmilling, in 1877 he and partner Henry Merrill moved to acquire control of a sawmill in McIndoe Falls, Vermont. McIndoe Falls was a favorable industrial site at the foot of the stretch in the Connecticut River known as the Fifteen Mile Falls and had supported a water-powered sawmill as early as 1816, along with a dam extending clear across the river.

The mill had been destroyed by fire in 1875, after which lum-

George Van Dyke.

bermen R. E. Peabody and E. E. Putnam bought the site and in the same year built one of the finest sawmills in New England at a cost of $50,000. This new mill employed 75 to 100 men and was capable of sawing 60,000 board feet per day from logs floated down the Connecticut.

Though the sawmill capacity was large, difficult times ensued at the McIndoe Falls operation, and the owners were forced to declare bankruptcy in February 1877, opening the way for a lease agreement and eventual purchase of the mill by Van Dyke and Merrill. Production at the mill leveled off at about 15 million board feet of softwood per year, most of it planed at the mill. This made the McIndoe Falls operation the largest lumber manufacturing site in northern Vermont at the time. Van Dyke was prospering with his sawmill enterprise but still worked best when not encumbered by business partners. Within a few short years he bought out his sawmilling partners and exercised sole control.

George Van Dyke's expanding influence on the Connecticut River was bound to create a conflict with the major log-driving company on the river, the Connecticut River Lumber Co. All parties using the river as an avenue of log transport would obviously want to use the river channel at about the same time of year; problems were inevitable. Not surprisingly, a merger of efforts took place in 1884, putting Van Dyke in the position of general manager of the CRL Co.

Van Dyke was not one who could rest comfortably while being subservient to superior authority, however, and within two years a further merger of assets placed him behind the desk of the company president. His assets enabled him to receive part ownership of the extensive company timberlands (possibly 240,000 acres), the sawmills, and the personal property of the Connecticut River Lumber Co. and its sawmill-operating subsidiary, the Connecticut River Manufacturing Co. The annual production of the large Mount Tom sawmill in Northampton was by then fluctuating between 40 and 80 million board feet. In 1891 the Connecticut River Manufacturing Company was formed to own and operate the timber-harvesting, log-driving and sawmill operations.

In 1897 Van Dyke purchased sole ownership of the company from the majority owners, George Scott and Thomas Pearsall, paying them $1.5 million. He reorganized the company as the Connecticut Valley Lumber Co. in 1902. His reign as the Lumber King on the Connecticut River went undisputed, and the acronym CVL became a part of North Country lingo when referring to lumbering along the Connecticut River.

Despite his prominent position in the industry, Van Dyke did not detach himself from his hard-working men and their toils. His bulky girth was often to be seen out on the log jams, and it was difficult not to hear him bellowing out various orders in a loud and commanding voice. He was hard on the men, but rarely asked of them what he wouldn't have pitched in and done himself.

There's an expression that the romantic writers of the river drives have attributed to Van Dyke. When a man fell from the logs into the river, Van Dyke would supposedly bellow "To hell with the man! Save the peavey!" River drivers, he felt, were expendable. There's probably a certain amount of factualness to that claim, but a harsh judgment is hardly warranted. Van Dyke hired the best of the trade, paid and fed them well, and always had his lumberjacks and river drivers returning season after season, hardly the record one would expect from an abusive and heartless task master. Almost all of the immortal names of famous Connecticut rivermen were longtime employees of George Van Dyke. The old-time timber barons were tough and demanding, but they recognized skills

and faithfulness and were accordingly fair with their men, receiving their loyalty in return.

Van Dyke's employees, such as this log crew at Bloomfield, Vermont, in 1890, were loyal to their employer.

An example of Van Dyke's regard for his men is the long told story that many a woodsmen, while on an extended drunk in Boston and finding themselves flat broke, would go into Van Dyke's Boston office and receive a hand-out to tide them over and provide car fare back to the woods.

However, pity the outsider who stoked Van Dyke's ire. One local newspaper reported an incident in which a man named True Loverin of Nashua, New Hampshire, angered Van Dyke, whereupon he raised his horsewhip and laid it vigorously on Loverin's head on the street in Canaan, Vermont.

For many years Van Dyke would accompany the log drive behind a pair of high-stepping matched black geldings and he and the horses became recognizable figures all the way down the

George Van Dyke can be seen sitting in the back seat of his sedan while parked in front of Fred Robie's store in Bloomfield, in 1907. Van Dyke had a field office in the left side of the building.

Connecticut. Later, in 1904, he purchased a Stanley Steamer and hired young Fred "Shorty" Hodgdon as his chauffeur to race him up and down the river.

It's been said that Van Dyke had a remarkable ability to remember the men and horses that had toiled for him over the years. Consider this story from historian Robert Pike, the accuracy or factualness which, unfortunately, cannot be vouched for. In the fall of 1901, Van Dyke was in New York City on business and was observing a pair of black horses hitched to an express wagon. He stopped short and explained, "Those horses were at Charlie Layton's camp on Teapot Brook and were loaded on the horse raft at the Narrows that spring. Look at their feet." A check of the front hoofs revealed that they did indeed have the identification mark that the company made on all their horses.

Even in his commanding position as head of the largest lumber company ever to operate in the Connecticut River valley, Van Dyke continued to expand into other ventures. In 1887–88 he constructed the Upper Coos Railroad, a common carrier railroad extending from North Stratford, New Hampshire, to Beecher Falls, Vermont, and served as its president before it was leased to the Maine Central Railroad. The rails of the Upper Coos Railroad were extended by Van Dyke into Canada to form the Upper Coos and Hereford Railroad, an important segment which linked with a Canadian-built railroad to form a through line to Sherbrooke, Quebec.

The Upper Coos Railroad was built in 1887, using railroad ties that C. H. Weeks prepared at his mill about a mile up Keyes Brook in Canaan. The ties were floated down the Connecticut River to North Stratford. The river was so low that fall that the gates at First Connecticut Lake had to be opened to provide enough water. In November of 1897, Van Dyke drove the first rail spike himself;

his first two blows, however, struck only the rail, but he finally did manage to drive the spike home on subsequent attempts.

With the building of this railroad an important access was also provided for Van Dyke's lumbermen to more readily reach the timberlands of the CRL Co; many miles of difficult sled-toting of camp supplies to his loggers were suddenly eliminated. When the company logged the Indian Stream country in New Hampshire, supplies were railroaded to Malvina, Quebec, and toted the short distance over the border through the woods to the large Depot Camp established on Indian Stream.

A seven-mile logging railroad was also built by Van Dyke up along the east side of Hall Stream, on the United States side, to access more of the company timber. The logs were hauled by railroad down to Arlin's Siding, two-and-a-half miles north of Colebrook, where they were dumped into the Connecticut River.

George Van Dyke had two brothers, Philo and Tom, who worked for him, but not as equal partners. Philo began working for George in 1875 on the Hall Stream log drive, and when George purchased the sawmill at McIndoe Falls in 1877, Philo moved there to take charge of the mill and the river driving. He continued to work for the company after George's death in 1909, but was reportedly a difficult man to work under. One old-timer recalled how Philo would keep infuriating the men at pay time until one struck him,

Oxen were part of Van Dyke's stable. Driver Sam Keeble has two yoke of oxen on a large load of cordwood in front of the Canaan House.

Company horses were also stabled in a large barn behind the West Stewartstown office building. The large wagon behind six horses was used to supply camps on the log drive.

then deducted $10.00 for assault. Tom, George Van Dyke's other brother, resided in Hereford, Quebec.

George Van Dyke never married, although he developed a friendship in later years with an old-maid school teacher in Colebrook. It was strictly a platonic relationship, or so it was claimed. His greatest weakness, according to Robert Pike, was stock market gambling, which reportedly cost him much of his accumulated wealth.

As with most timber barons, Van Dyke loved and respected his horses. Besides owning homes in Hereford, Quebec, and in Colebrook, he also had some fine private stables in Canaan, Vermont. These housed up to sixty well-bred racehorses and a private racetrack. Some of pacers housed there were very well known in New England racing circles. There were also four large barns with over 150 head of cattle, all tended by Henry Knapp.

For twenty-six years George Van Dyke was personally on the Connecticut River, watching over his winters' investment of logs as they bumped and pushed their course southward to the mills. Van Dyke died tragically in 1909, tumbling over a cliff in his automobile at Turners Falls, Massachusetts, while watching his rivermen at work down below. He was sixty-three years of age.

The massive log drives from the Connecticut headwaters down to Massachusetts, of which he was such a vital part, lasted for forty-seven years (1868–1915), thirty-seven of them under the ownership of the Connecticut River Lumber Co. and the Connecticut Valley Lumber Co.

The greatest and most memorable of these log drives, as well as the ones photographed, were the drives conducted by the Connecticut Valley Lumber Co. at the turn of the century. Many memories, incidents, factual data and photos have been passed down to make it possible to document this exciting chapter in logging history.

The following section of this book will relate the lumbering activities of this era, particularly as they concern the CVL Co. Starting with a description of logging the old-growth timber, the account will then follow the spring log drive, starting at the very top of the Connecticut River and ending at Mount Tom, Massachusetts.

7

Harvesting the Old-Growth Timber

WINTER TIME was a busy season in the North Country, for this was when the yearly supply of logs had to be cut. Once felled by the lumberjacks' trusty axes, the logs then had to be pulled out to a location where they could be piled and ready for coaxing into the river current when the spring "break-up" occurred. Typically, George Van Dyke hired 600 to 700 loggers, some 600 horses, plus the employees of the independent jobbers, to work through the long and cold laborious fall and winter seasons toiling for the Connecticut Valley Lumber Co. The old-growth timber in the woods making up the headwater of the Connecticut River was primarily made up of spruce and a much lesser amount of balsam fir, and that's what the loggers were after. White pine was also cut when encountered, as was hemlock on occasion. Thus, all of the merchantable conifers or "softwoods," except cedar and tamarack, were sent down in the Connecticut River log drives.

The men would enter the woods in October or November and remain there until the start of the log drive in March or April. While in camp throughout the cold winter months, it's been said that many of the woodsmen rarely, if ever, bathed. Some, according to their contemporaries, never even changed their underwear all winter long.

With so many men in the woods during the winter season the logging camp had to be an indispensable part of the scene. The woodsmen had to be kept alive and healthy for six months and contented enough to keep them from wandering. At the same

All of the merchantable softwood was harvested. This photo shows the Paul Stream area, about 1911.

A logging camp constructed of logs set vertically.

time, camp construction costs had to be kept to a minimum because once the area was logged off the camp was abandoned.

During the earliest years of logging in the Connecticut River headwaters, the logging camps were crude, low one-story buildings of log construction. A visitor to the Connecticut Lakes in 1896 described one CVL camp as having two rooms, a 20- by 30-foot room for the workers and an 18- by 20-foot room for cooking and dining. The roof had originally been covered with two layers of wood splits, the first layer flat side up and the top layer flat side down covering the joints. The floors had once consisted of small poles, hewn on top, but both the floor and roof were later covered with boards and tarpaper.

A CVL Co. logging camp.

Bunks were frequently a single, long platform with straw and fir boughs for a bed and a single large heavy covering over the men laying side by side. Heat for cooking and for warmth might be supplied by an open fire in the center of the cabin, vented through a hole in the roof.

During the latter 1800s, however, the logging camp scene became more tolerable. The buildings were log or wood framed in construction, boarded over and covered with a couple layers of tarpaper, and they became tall enough for a husky logger to stand

The CVL Co. used log construction for many of their camps, but there was probably more than one bunkhouse for these approximately 95 men and one dog.

The cooking crew and camp boss pose in front of the cooking and dining building at a camp on Paul Stream. Courtesy of Lester Fogg.

once inside. The bunks became single beds, frequently tiered two or three high, and the men had their own blankets. A larger camp could even have a two-story bunk house with maybe a dozen rooms on the lower floor for the teamsters and enough beds for fifty men on the upper level.

The dining room and kitchen were normally located in a building separate from the bunk quarters. On the kitchen end would be

The cook and his helpers prepare an abundance of hearty food for the loggers at the CVL Co. camp on Paul Stream. Courtesy of Lester Fogg.

found a couple of wood stoves, a flour bin, pastry drawers, a large variety of pots and kettles, food supplies, and the wood dingle to hold the firewood. The long tables in the mess hall might seat as many as fifty men at each table.

The meals were plain with little variety, but quite wholesome and savory and loaded with calories. Breakfast would typically consist of hot biscuits, sweet bread, doughnuts, dried applesauce and baked beans. Everything was sweetened with black molasses, and strong, dark tea was always on the table, but never coffee.

Potatoes, pastries and pork and beans were the steady basis of most meals. The cook made his own bread, pies, cookies, doughnuts and cakes, and with most of the camp cooks these pastry items were a delicacy.

The crews working near the camp would come in for lunch, but for those working in distant locations one of the cookees had the chore of hauling lunch to them in buckets slung onto a neck yoke placed across his shoulders. He would build a fire to make the hot tea and thaw out the food. The men might find a seat while eating, but if it was quite cold, they would eat standing up and quickly hurry off.

Supper was served at some point after dark, usually about 5 P.M.; no daylight hours were wasted for the crews in the woods. No matter where one traveled there was always the universal rule that no talking was permitted at the dining table during the meal. The only sound was the clatter of knives and forks on the tin plates, the musical effects of liquids being swallowed, and the snorts and grunts of satisfaction. The cook still had work to do after the meal, and he wanted the men fed promptly and up and out when finished with the meal.

A good cook was a necessity and was treated with great respect. The loggers expected good food; if it was bad, they might quickly pack up and head down the road to another camp. Among the cooks prized by the CVL camp bosses along the Connecticut were Tom Brackett, Fred Brooks, Dan Kelliher and Joe Buckshot.

Transporting the supplies to the camps was the job for the tote teamster. With a rugged platform sled and with 2 to 8 horses in front, he obviously had to deliver regardless of road conditions, in good weather or bad, over roads mired in mud or deeply rutted or buried in snowdrifts, and across frozen lakes with the ever present danger of a break-through. Be it a blizzard or an unseasonal thaw, the tote sled had to deliver to keep the camp operating. Without any snow on the ground a dray or "jumper" would be used. This was a platform mounted on a couple of poles fastened to a pair of high wagon wheels. If the rig became stuck in a mud hole the teamster would have to unload all of the supplies so that the horses could pull it free.

Daily Food Needs for One Camp

The daily food consumption could be prodigious for lumberjacks. Dan Kelliher was cooking for 100 men in the winter of 1918–19 at the Connecticut Valley Lumber Co. camp on Scott Bog, above Second Connecticut Lake. The daily consumption was as follows: 75–100 lbs. beef, 3 bushels potatoes, 2 gal. tomatoes, 40 lbs. sausage, 25 lbs. liver, 30 pies (apple, mince, cherry, raisin, lemon, prune), 200 doughnuts, a bushel of cookies, 20 one-pound cans of condensed milk, 10 yeast cakes, 3 gal. canned apples, 16–20 large double loaves of bread, 2 gal. molasses, cabbages, turnip, tea, oatmeal, plenty of beans.

At W. Reed's camp for the Nulhegan Lumber Co. in 1887, cook Charlie Jacobs was feeding 70 men. The average amount cooked daily was reported as: 12 quarts of beans, 2 bushels potatoes, 430 biscuits, 339 doughnuts, 230 pieces gingerbread, 19 pies, roast beef, corned beef, fish, and on Sunday, a pudding that would fill 2 bushel bags.

After enjoying a hearty lunch, teamsters head out to the log landings again to pick up the sleds.

A tote team brings in a load of hay to the CVL Co. camp at Brown's Mill on Paul Stream, where the company maintained a storage barn. A spring pole was used to hold up the pole so that it would not weigh on the necks of the pole horses. Courtesy of Lester Fogg.

The cookee (or cook's helper) would rise by 4 A.M. or even earlier, build the fires, and call the cook about a half-hour later, being careful not to wake the men quite yet. By 5 A.M. breakfast was ready, and the men were rousted by the cookee clanging the gut hammer. After washing their hands and faces in a pan of cold water, the men then headed for the mess hall.

A supply sled makes its way to camp.

> ### *The Dirtiest Cook in the Woods*
>
> Joe Bouley (Joe Bully) from Trois Riviere, Quebec, was considered to be the dirtiest cook along the Connecticut River valley. His cooking was satisfactory, but the cook room became so filthy and greasy that it was said to require caulked boots to safely walk through. One camp boss stated that Joe was so dirty that you couldn't hang onto him with a pair of sharp ice-tongs.
>
> In 1907, woods boss George Yates was opening up a new camp on Paul Stream and had to fire Boulay when the men complained about the food. Van Dyke sent Boulay from Bloomfield, Vermont, back into another camp to work as the cookee. Van Dyke said it took a tub of lard and a half-barrel of flour to get him greased up for the winter, and he wouldn't do it twice in one season.
>
> On another occasion, scaler Rube Hall recalled how Boulay rigged up a rat trap over a barrel of molasses. He arranged a teeter board over the open barrel of molasses with the bait at the end of the board so that the board would tilt and dump the rat into the molasses. In the morning Boulay would fish the dead rat out of the camp's molasses supply.

The teamsters would be roused by the cookee a half-hour or so before the other workers, and usually at least an hour before daylight, because they had the additional chore of feeding their horses before breakfast. If there were icy sled roads on the slopes that needed to have hay or sand spread on them to help slow down the loaded log sleds, the "roadmen" had to be up and on the job during the dark hours of very early morning, before the teamsters began their hauls.

The camp clerk slept in the little building that also served as the camp store, where any necessary items, from clothing to cut plugs of tobacco, were readily available to the camp crew. The clerk was responsible, too, for the time cards kept to account for each man's labor. The camp boss usually slept in this little office, too, as well as the visiting walking boss or the scaler.

Daylight would have hardly broken forth before some of the cutters were on the scene felling and limbing the spruce, letting "light into the swamp" as they were wont to say. During the mid-1800s the trees were still felled, limbed, and bucked into log length by means of axe alone. By the latter part of the century, however, the two-man crosscut saw had found popular acceptance for felling and bucking. Contractor Wilbur Marshall's first use of the crosscut saw by his crews was in the winter of 1902–03 at a camp on Gadwah Brook. It was also at this camp that Marshall's cooks first used bread tins instead of having to just rely on the baking of biscuits.

Clothing worn in the woods and on the log drive was functional and somewhat unique, but was seldom fashionable. The underwear was heavy woolen, seldom washed throughout the winter. Dark woolen breeches were common as were heavy woolen shirts with tails on the outside, frequently red in color. If cold enough, an

The clerk's office also housed the camp store and a bunk for the camp boss. The location of this office is the Bog Dam camp on Paul Stream in 1908. Courtesy of Barney Hall.

extra shirt or a mackinaw was worn. When on the log drive, the trousers were fastened inside the caulked boots; in the woods the men commonly wore rubber footwear with long woolen socks and loose trouser bottoms.

Injuries on the job received little, if any, attention and certainly very little sympathy. The hospitalization of injured workers was

A camp cook, pictured with his wife and children, takes pause with other loggers to have their photograph taken at this CVL Co. camp. Two of the loggers seen below are believed to be Ivan Willis and James Davis.

rare, but loggers with serious wounds were somehow eventually dragged down to see the village doctor. If the woodsmen died or had drowned, he was, after all, only a logger, or so went the thinking back then.

Lumberjack Marsh Underwood, who worked along the Connecticut in the early 1900s, told in his published story about the time he had turned his foot and hurt it badly. He hung around the bunkhouse for a couple of days, without any treatment, and suffered plenty of razzing by his fellow workers. Finally, he could stand it no longer and somehow pulled his boots on despite the swelling and limped out to work.

Cleanliness among the woods workers was obviously not a priority either. In January 1880 the *Essex County Herald* reported that measles was raging in the lumber camps in Pittsburg.

HORSES

The woods horse was the essential workmate and companion of the logger from the very start of the tree harvest through to the final delivery of the logs at the sawmill down river. Horses were even active participants during the river drives.

The best horse for woods work was between 5 and 10 years old, weighing between 1,300 and 1,800 pounds. Buyers looked for horses with legs and neck short and stocky, and with well-developed muscles in the shoulders and hips.

A quiet and unexcitable temperament was very important too; excitable and nervous horses could not pull as heavy a load. But even the best-tempered horse could be spoiled by improper treatment. Most teamsters, therefore, treated their horses firmly but kindly; the few who didn't probably didn't stay around for long.

A logging crew and skidding team as seen along Back Lake Road in Pittsburg. Identified in the group are, from left: Harvey Day, Jack Chappell, Simeon Scott (scaler), Daniel Heath, Rodney Scott, Gilbert Scott, and the boys Austin and Fred Scott.

The roads were kept well iced in the sled ruts on the level or on an upgrade to ease the burden on the horses with a heavy load such as this one.

The loggers appreciated their dependency on the woods horse and usually gave the horse a considerate care that exceeded that given to the men by the logging boss.

During the late 1800s most of the horses were Belgians, often obtained from farms in Ontario or from dealers in Montreal, Quebec, for a price in the vicinity of $80 to $110 each. Will Fuller was one of the men who had charge of the woods horses for the CVL.

A good team horse knew the voice and special commands of his teamster and was able to quickly sense various hazards. A well-trained skidding horse could even work alone without an escort, twitching a single log from the cutting crew down to the man at the skidway and returning without anyone walking along with him. If the log he was dragging was to hang up en route on an obstruction such as a rock or a stump, the horse would know enough to "gee and haw" in different directions on his own until the log came free.

If a horse fell in deep snow it could become quite difficult for him to get back on his feet with the harness still in place. A trained horse would know enough to lie still until the teamster had unhooked the straps and chains. Working eleven hours a day during the logging season, a horse could be expected to last about six years as a woods horse.

Work in the woods was hazardous for the horse as well as the men, with deaths occasionally occurring when a tree fell on the horse or when a team was "sluiced" (or overtaken by a heavily loaded sled out of control) on a steep hill. Veterinarian Edwin Blaisdell of Haverhill, New Hampshire, told of a woods horse he was asked to treat after it had been hit on the rump by a large spruce tree. The horse was so tall that Blaisdell had to stand up on a truck body in order to look down into the wound and treat it. He then removed a red bandana that the teamster had stuffed inside to stop the bleeding. Returning a couple of nights later when the horse was hitched in the hovel, Blaisdell climbed up onto the rafters to enable him to see the wound, and with the aid of a flashlight noticed a bit of bark sticking out of it. With the aid of a large pair of pliers, Blaisdell gave a yank, and as the horse gave a mighty

The Sluicing of Old Ginseng

The following is a condensed version of an event described by Vern Davison in Robert Pike's book, *Spiked Boots.*

On Cowen Hill, up along Indian Stream, there was a steep hill near Graham's Camp that required loaded sleds be snubbed down for about 800 feet with a strong rope. On one occasion a teamster named Ginseng had 5,000 board feet of 20-foot logs fastened down on his sled by three binder chains, and he started down the slope while hooked onto the 2-inch manila snub warp. The snubber at the top, a small French-Canadian, gradually let out the rope, which was customarily wrapped around the stump in three coils. To further slow down the descent there was a wooden lever set in a mortise, which could quickly tighten up the rope as it was played out.

Ginseng had gone about 250 feet when the snubber carelessly bore down too heavily on the lever, and the slack was taken up so quickly that the sled and horses were actually jerked back up the hill a little. Ginseng, standing on the front end of the logs, shot out over the horses and slid to a stop on his face and stomach along the sled road. The snubber obligingly held the rope tight to allow Ginseng back on the load.

But Ginseng's troubles had only just begun. As the team eased the load over the next hogback above another drop-off, the rope parted at a point that had just been weakened by the snubber's previous misdeed. The sled started down the steep slope with no way to brake the accelerating descent and no way to guide the team off the road because of ledges and trees on both sides. Riding on top of the reeling logs, Ginseng hesitated to jump at first, then realized it was too late anyhow because of the increased speed. As the lurching sled hit another bench in the road the logs shifted under the tight chains and a log rolled over on top of Ginseng's foot. He quickly pulled his bare foot out of the trapped boot.

Shooting down the next pitch, the horses had no way of holding back the careening sled behind them. They were not actually running, just jumping, and every time their feet would come down the load would jam them forward into another leap. Ginseng was riding it out because the horses had so far been able to keep on their feet.

The road wasn't straight, however, and when they hit the last steep pitch, Ginseng was unable to swing the horses around the next curve. The momentum of the loaded sled pushed the team off the road and hurled them against a large uprooted leaning spruce tree. Coming to a sudden stop there was a horse on each side of the leaning tree, hung up there by the neck yoke with only their hind feet on the ground.

Ginseng, meanwhile, had jumped unhurt into a snow bank before the sled hit, and he quickly ran ahead to check on the welfare of his horses. Grabbing an axe, he climbed up on the leaning tree and quickly cut the neck yoke before the horses choked. Ginseng then headed back up the hill on foot, axe in hand and cursing loudly. The snubber saw him coming and reportedly ran like a fox for parts unknown.

lunge, the vet pulled out a piece of spruce branch twice as thick as a broomstick and over two feet long. The wound eventually healed.

SLEDDING

From the stump where the tree was dropped and cut up, the logs were moved down to a gathering point where they could be loaded onto a sled for the long journey down to the edge of the river or a lake. Moving the log to the skidway where the sleds were loaded might be by skidding or twitching a single log behind one horse if the distance was short. Or it might be by use of a bobsled pulled by a team. The bobsled was a single log bunk mounted over a set of runners. The front end of a small bunch of logs could be chained on top of the bunk and the back end of the logs would be allowed to drag on the ground. Bobsleds were especially effective on slopes because the dragging logs acted as a brake when descending a steep slope.

A long, steep downgrade might require an extra measure to prevent the load of logs from overtaking the fast-stepping horses and sluicing them off the sled road, possibly killing them. In this situation the logger would snub the load down by fastening a rope to the rear of the sled, wrapping the other end around a stump at the top of the slope where a helper would play out the rope to check the sled speed. There weren't many long, steep mountain slopes in

The small end of the log was usually the end placed forward onto the bunk of the bob sled for better control.

The bob sled or yarding sled consisted of a single bunk mounted on a set of runners. This photo was taken in the Paul Stream area. Courtesy of Lester Fogg.

Teamster Frank Gendreau drives a two-sled with a load of 3500 board feet of spruce log into Camp 2 in the Paul Stream country. The year was 1908 and the longest logs cut by the lumberjacks were 36 feet. For Gendreau and his team, it was three-mile sled haul to the stream where he would unload the logs on the frozen surface.

the CVL Company's North Country timberlands, but in 1884 the company was logging off Monadnock Mountain in Lemington, Vermont, and was snubbing the sleds down the slope with 7,000 feet of 1½- and 2-inch rope.

As with many of the logging duties, snubbing could be dangerous for the careless worker. Lurvey Mayo, a logger from Sherbrooke, Quebec, had an unfortunate accident on a CVL job in Bloomfield, Vermont, in January of 1911. Not realizing that he was somehow entangled in the snub rope, his foot was reportedly taken right off when the rope was suddenly yanked taut by the team.

The sled type commonly employed for long hauls was the two-sled, actually two independent bunks, each mounted above a set of runners, one bunk under each end of the log load. The heavy weight of the log load chained to the bunk held the bunks in place,

Teamster Joe Burke handles a well loaded two-sled on Paul Stream. Joe was said to be one of Van Dyke's favorite teamsters and was thus given the best horses. A false pole can be noted between the horses in front, which allowed the lead team to also hold back on a down grade. Courtesy of Lester Fogg.

but underneath the load there was a set of chains strung between the sled bunks. The chains were crossed over and hitched to opposite corners so that when the front bunk sled made a turn the rear sled would be swung over in the opposite direction to accommodate the curve in the road. The bunks rested on a swivel that allowed the runners to move within the curves of the ruts in the sled road.

Loading logs onto the two-sled demanded coordination and skill by the landing crew. From an elevated skidway supported by a cribwork of logs, the long spruce logs were rolled down and onto the sled. The first log was placed in the middle for balance, and then a log was chained on each side and other logs packed in between and above. To unload the sled, the lids hook in the bind-

Pictured here is a large yarding area in the Pittsburg area where two sleds were loaded.

Woodsmen load long spruce logs onto a two-sled near Paul Stream in 1906.

ing chain would be knocked loose, usually allowing some of the logs to roll off on their own without prying because the sled would be stopped in a tilted position if at all possible.

Skidways were sometimes built at a height that allowed the crew to build a high load on the sled, especially when the teamsters were contesting to determine who could draw out the largest load, as

The measured volume is unknown but the load of logs seen here was said to be the largest ever hauled out of the Paul Stream drainage. Teamster Peter Petegan was probably given some selected logs for this feat in 1911.

Pictured in this 1906 camp photo are the "bull punchers" and their bulls or oxen, which were used on a few jobs.

they occasionally did. Such was the boast of teamster Thomas O'Brien in March 1880 when his pair of horses, weighing 2,900 lbs., drew a load of 4,650 board feet of spruce logs for one-and-a-quarter miles up on Nash Stream. Testing the limits of credibility was the 1902 claim by a logging camp up on Paul Stream that they had put together a sled load of selected 20-foot spruce logs measuring 16,000 board feet. Perhaps a bit more credible is the mark reportedly set by the crews at Granby Camp on Paul Stream. In 1913, they supposedly cut and hauled out 600,000 board feet of logs in a single week.

Sledding couldn't begin for the season until there was a good amount of snow on the ground and the road crews were able to pack down a good road surface. Water was spread on the sled road at night to build up ice in the sled ruts. But without snow, sledding was at a standstill, and there were occasional winters when the weather let the loggers down. For instance, in mid-January 1889 the Nulhegan Lumber Co. was reported to have 100 horses and 250 woodsmen idled at their camps in Essex County, Vermont, because of lack of snow.

WOODSMEN

The workers in a logging camp were classed by the specialty of their work—chopper, teamster, peavey or yard man (log roller), road swamper, blacksmith, filer, etc. Many of the jobs required skill

Some Woodsmen Fought Back in Court

Court cases arising from injuries suffered in the woods were frequently filed against Van Dyke or the CVL Co. Among them were the following.

1904: Nine Canadians were hired through employment agent Robert St. George to work as lumberjacks for the CVL Co. near North Stratford, New Hampshire, and were provided with railroad tickets to travel there. When they later went to collect their wages on payday, each was compelled to enter the office alone and there they were confronted with agents armed with revolvers who forced them to accept much less pay than had been agreed upon. The case was heard before the Essex County Court in Vermont, which determined the back pay due the men.

1910: Joseph Valley was attempting to thaw some frozen dynamite before an open fire on a log job in Ferdinand, Vermont. The dynamite began to burn and Valley picked it up to move it despite being warned not to do so. The charge exploded and Valley lay there for six hours suffering. He died later. His estate asked for $10,000 in damages, claiming that Valley was inexperienced and that the company failed to provide proper supervision. Testimony showed that Valley did have previous experience with dynamite and the claim was denied.

1914: A similar suit was brought by John Cachuke of Berlin, New Hampshire, who was blasting out some rocks while constructing roads for the CVL Co. He lost his right eye during a premature explosion and claimed the company failed to instruct him in the proper use of explosives and to warn him of the dangers.

1918: Nineteen-year-old John Paquette was hired to fell trees and given an axe, saw and sledge. While cutting a large tree he asked the boss, a Mr. McDonald, for an iron wedge but was given one made from a dry spruce limb. When he struck the wedge a piece of wood flew off and severely injured his eye. It was revealed during the court case that he had never cut timber before.

and experience, and the loggers took pride in their respective "calling." Some would not admit that another man could possibly be a match for their abilities.

There were always some loggers who tended to drift from camp to camp of various operations, always discontented. "Camp inspectors" they were called. George Van Dyke reputedly paid well enough, and the drifting was kept to a minimum in his camps. His brother, Philo, became quite intimidating on payday, however. He had a habit of making pay deductions for inappropriate reasons, and rumor has long had it that he strapped a pistol on his waist when paying the men. Wages for the loggers varied from $15 to $25 per month, plus room and board, and didn't change much until the First World War created a manpower shortage in the woods. The company's boarding cost was $2.00 to $2.50 per week for each man.

There was little on which the logger could spend his money while in the camps, and almost all chose to wait until the end of the season to collect their pay. Sometimes they had no choice. They

Clothing didn't vary much among the loggers, as seen in this photograph taken at Charles Knight's camp on the Nulhegan River. Knight was a camp boss for the CVL Co.

The Roby Family

The Roby men became renowned among the logging fraternity along the Connecticut River. The father, Joe Roby, Sr., came from Paquetville, Quebec, and raised 8 boys and 8 girls. His wife had died from blood poisoning resulting from an axe wound in her foot.

The best known of the boys were as follows.

JOE ROBY: Had a farm at Pittsburg, New Hampshire, and worked as a walking boss for Van Dyke at a number of locations. He was liked by the French Canadian woodsmen and as a result the CVL Co. named some camps along the Nulhegan River as "Joe Roby No. 1," "No. 2," and "No. 3" to keep the French Canadian loggers from going elsewhere when they arrived by train at North Stratford, New Hampshire. He died in 1919.

CHARLES ROBY: Started working in the woods at age 10 driving a team of two oxen. At 19 he was foreman of a lumber camp and soon became walking boss for Van Dyke at the camps and on the log drive. Charles later worked for other operators after the log drives were finished, working 59 winters in the woods without a break before his death in 1934.

ALPHONSE (PHONSE) ROBY: Started working with oxen in the woods at age 14. He eventually became a walking boss for Van Dyke at the lumber camps and the log drives. Phonse wasn't much of a talker, but he saved the CVL Co. so much money by being able to handle the log drives through Fifteen Mile Falls that he was made the walking boss for all of the CVL timber operations. He died in 1942.

The other brothers (Lawrence, Fred, Henry, Frank and Barney) all worked for George Van Dyke on his Connecticut River log drives at some point in their lives.

Methods of Landing the Logs Streamside

There were three general methods in "landing" or placing the logs by the water preparatory for river driving.

1. Logs were piled on the riverbanks in rollways. A high bank was chosen where possible so that the logs could be rolled down to water's edge onto a pile. As the river rose in the spring the pile would often be flushed automatically by the water. If a meadow or low bank location had to be used, the logs were piled high by means of a donkey engine and winch and later rolled into the river by means of the engine placed on a raft in the river.
2. Logs were piled on the frozen surface of the stream from bank to bank with the lengths parallel to the stream flow. They were piled high and might fill the stream surface for a half-mile or more with about 20 feet between piles. The logs over the current were freed first when the ice let go, then other logs would begin to roll into the current until the whole pile "hauled."
3. On the larger streams or rivers the logs might be piled on the ice, parallel to the current, along each side of the stream. The center of the stream was kept open as a roadway for the sleds. The piles would "haul" on their own with the ice break-up.

A high bank near Granby Dam on Paul Stream was used to build this rollway in 1895. The steam donkey and pulley on the A frame (upper right) were used to pile the logs as the sled brought them in from above. Courtesy of Lester Fogg.

Paul Stream is filled with spruce logs awaiting the spring ice break up in April 1914. This method of preparing logs for transport down river was used on smaller streams. Stone and Webster report, 1914.

might keep a tally of days worked by a record in an account book or by making notches on a stick hidden in the bunk. As best they could, a tally was kept of supplies purchased at the camp store, items such as chewing tobacco, a new pair of rubbers or an article of clothing.

Sunday was the only day work was suspended, short of a severe storm. Men would relax on Sundays, mending clothes, bathing, playing cards, repairing axes, hunting rabbits or hiking to town if close enough. But most of them stayed at the camp for at least four months without going outside, some even had 6- to 8-month hitches, and the occasional individual stayed right in the woods for one to two years.

More than anyone else, the men who carried the greatest responsibilities at a well-functioning log job were the bosses—the walking boss and the individual camp bosses. They laid out the sled roads and the camp locations. The camp boss also made it his business to be on the job in the early morning, checking on the condition of the sled roads and the treatment of the horses.

The Connecticut Valley Lumber Co. had some reputable and well-known walking bosses. There were the brothers, Joe, Charles and Phonse Roby, Bill Smart, and Win "Grinner" Schoppe. Bill Smart had a group of PI's (Prince Edward Islanders) working for him on one operation; he considered them to be good woodsmen, but not very smart.

All of the logs cut by the men had to be measured, tallied, and marked by the CVL Co. scaler as they were brought down to the gathering point at the water's edge. The normal company mark on the log end was four X's placed in a diamond pattern. The purpose of branding logs was to identify ownership if the logs had to be sorted from other ownerships while in the river. Ed Bateman and Orton Newhall were two of the CVL scalers remembered years later. The scalers would record separate log tallies for each of the teamsters, and Bateman had what he called a "bucking board" posted on the camp wall to show the tallies of the best teamsters.

The Connecticut Valley Lumber Co. seldom clearcut the land of all tree cover. The hardwood trees were left untouched, and George Van Dyke had a standing order for his woods crews that no softwood tree should be cut if it could not yield a 20-foot-long log at least eight inches in diameter at the small end. The reason behind the order was strictly one of economics, not conservation, but young growth was left remaining. The company was logging 42 to 50 million board feet yearly and with an estimated 1.2 billion board feet of sawtimber still standing at the turn of the century on its 330,000 acres, the company reckoned the timber harvest would last another 30 years. It didn't.

Log piles along the river edges were used on the larger streams and rivers, as shown along the Connecticut River below North Stratford in April 1914. Note the rollways along the top of some of the piles to extend the pile out further on top of the ice. Stone and Webster report, 1914.

Stacking the logs on the edge of the river or piling them out on the ice surface brought to a completion the accomplishments of the winter season. The logs then awaited the spring "break up" when the currents flowed fast and the drive began.

The men took their winter's wages and wandered out of the woods, many to seek a wild spree of entertainment and in the process part with much of their hard-earned season's wages in a brief fling. The April 11, 1879, issue of the *Essex County Herald* reported that the streets of Canaan, Vermont, were thronged with men, drunk and brawling. The nearby Canadian Line House contributed to much of the drunkenness. Five hundred men just out of the winter woods and crazy for booze could stir up quite a dust.

The following year, the camps of the CRL Co. also closed down the first week in April, and within just a few days the fighting commenced again at the local bars. In one instance a terrible fight broke out at William Tibbitts' hotel in Pittsburg, resulting in the severe pounding of a Quebec logger, after which the participants moved on down the river to an establishment kept by James Bacon, where another severe fight soon broke out. Bacon's establishment, often referred to as the "slaughter house," had an open bar with the windows boarded up on the inside to prevent the combatants from throwing men through them. After beating a man named Pat McNulty, the parties continued on down the river to Canaan village. Two days later four of the men involved in the barroom assaults were arrested at West Stewartstown, and after one had escaped, they were taken to Colebrook and eventually fined.

Prejudice against foreign immigrants would surface especially during these drinking bouts. Logger Peter McKenney over-imbibed on a January spree in 1913 and ended up in a fight with a Russian woodsman. The Russian whipped out a knife, gave McKenney a three-inch long slash on the arm, and found himself in jail soon after.

Liquor, though not allowed in the lumber camps, still managed to make its way into camp on occasion. Logger Peter Poulin was shot and killed by Carl Dolley during a drunken spree in 1914 at a camp on the East Branch of the Nulhegan River, three miles above Stone Dam. Dolley eventually pleaded self-defense and was later cleared of the charges.

Not all the woodsmen took part in the spring revelry. There were a few who returned home at the season's end, back to their home farm for spring chores. The majority, however, would bide their time at the Connecticut River headwaters awaiting the ice break-up and commencement of the river drive. During the week following the brawl recounted above, newspaper accounts noted that there were 600 men gathered at First Connecticut Lake in anticipation of work on the drive.

At the conclusion of the winter woods work, the horses were walked out of the camps as soon as the hauling was completed, and the sick and lame ones were tended to at the company farm. The ones to be kept were put out to pasture for the summer at Tom Van Dyke's farm in Canada. Those not to be kept were taken to the big barn to have ailments treated and to exercise them until they could be sold. A few horses were also chosen to be taken down the Connecticut River on the tail end of the river drive for use in hauling stranded logs off gravel bars, ledges, and farm meadows.

With the winter logging season over and the logs all streamside in anticipation of a good spring log drive, the next phase of the woodsman's exciting yearly routine was about to begin.

8

The CVL Company Log Drive on the Upper Connecticut River

Connecticut Lakes Down to Woodsville

THE ADVENT OF THE ERA of the log drives on the Connecticut River opened a new way of life for the North Country. The occupation of a logger or a riverman became elevated to a major status in what had formerly been a quiet agricultural valley. The once quiet rural villages became transformed into a bustle of activity for the short season of the year when the lumbermen were in the area.

Establishments for lodging, drink and meal made their appearance to cater to the logger who had seen nothing but trees and snow for five winter months, or to the riverman passing down the river. Red light establishments sprang up in the major towns along the river, such as Woodsville and Bellows Falls, to cater to those loggers who preferred to skip the church socials.

North Stratford, New Hampshire, had been a small but active railroad village since 1850, before the era of the log drives; it became a booming community afterward, the log driving center for Vermont's Essex County and New Hampshire's Coos County. A half dozen hotels and boarding houses opened up, certainly not in keeping with the needs of the small local population. Taverns became plentiful, and the livery stables were busy during the seasons when horses were moving down or up the river. Once the log drive had passed, the shanty town with its comely women and other entice-

The ice has given way and the current is beginning to move a few logs down the Connecticut River in April 1914. But plenty of hard work was yet ahead to put all of the logs in the current. Stone & Webster report, 1914.

ments to lure away the logger's hard-earned money would disappear as quickly as did the tail end of the log drive; the village would once again settle down for a quiet summer before the loggers again made their fall appearance, headed once more for the woods.

The village boys, however, never forgot their heroes of the river who fearlessly rode the logs on down to destinations unknown to a country lad. A stray log or two left behind in the river gave them stage to act out their fondest dream; they wanted to grow up to be a river driver, to proudly stand on a big spruce log as it careened down through the whitewater to another world down below.

Anticipation in the river villages began with the first signs of winter's end, and no time of the year was more eagerly awaited by the woodsman. As the ice began to break apart in certain areas, the call went out for all the river drivers to be ready. And many good men were available. To be one of the large river crew going down the Connecticut with the annual log drive was a privilege and an honor, something to boast about to the next generation or two. Long days of hard work, standing in cold water for many hours a day, discomforts and extreme danger all headed the list on the job description; respect and fame were the rewards.

Thus it was that the break-up of the frozen streams was an event fervently anticipated, especially since most of the river drivers had just spent the winter cooped up in a logging camp deep in the woods. When the sled roads deteriorated into deep mud as the spring thaw transformed the frozen woods, the seasonal logging jobs were done. Any logs still in the woods would probably remain there until the next season.

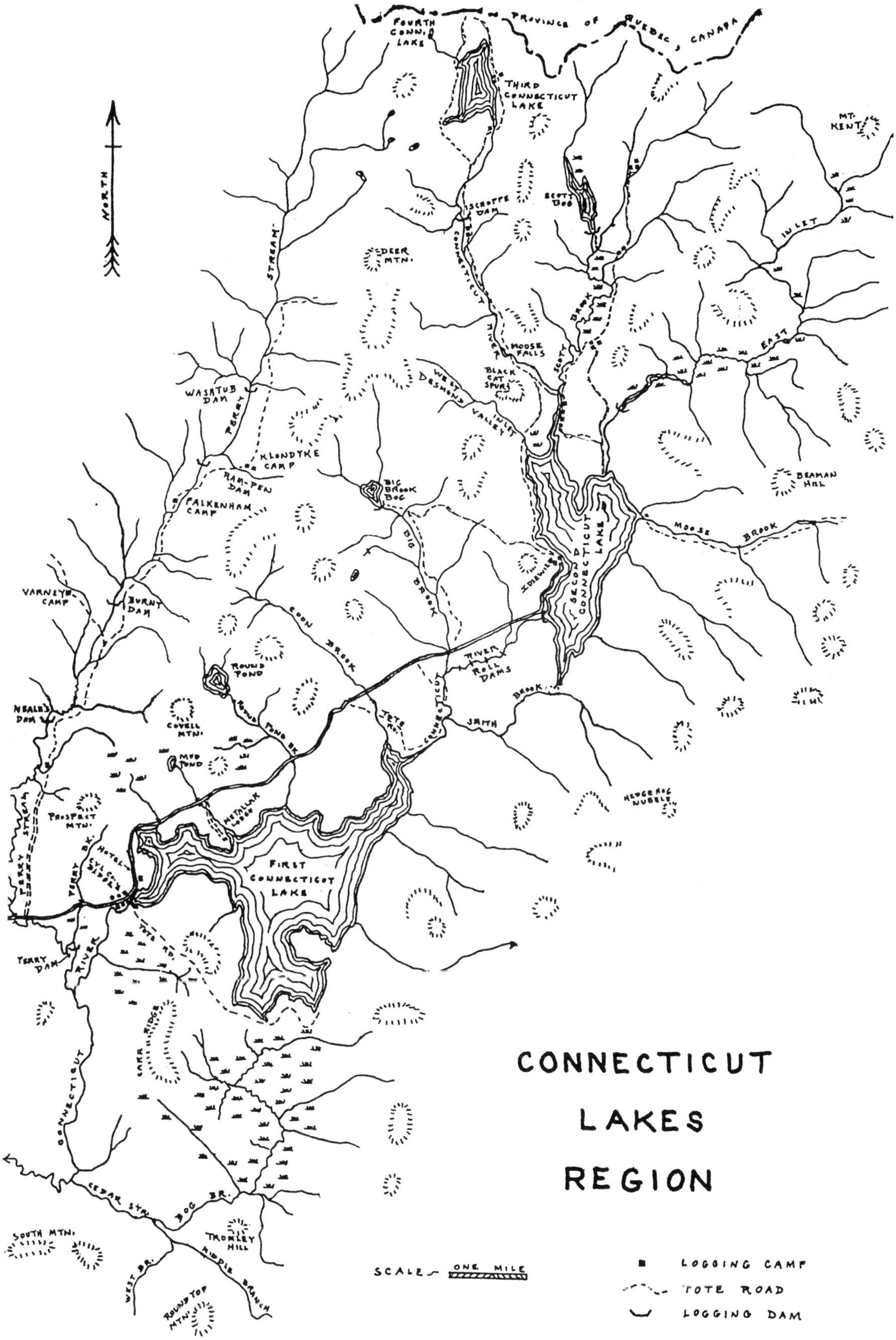
PROVINCE OF QUEBEC, CANADA
NORTH
FOURTH CONN. LAKE
THIRD CONNECTICUT LAKE
MT. KENT
SCOTT BOG
SCHOTTE DAM
DEER MTN.
PERRY STREAM
CONNECTICUT RIVER
INLET
EAST
MOOSE FALLS
BLACK CAT SPURS
WEST DESMONS VALLEY INLET
SCOTT BROOK
WASATUB DAM
KLONDYKE CAMP
RAM-PEN DAM
FALKENHAM CAMP
BIG BROOK BOG
BIG BROOK
BEAMAN HILL
MOOSE BROOK
SECOND CONNECTICUT LAKE
IDLEWILD
VARNEY'S CAMP
BURNT DAM
COON BROOK
ROUND POND
RIVER ROLL DAMS
BROOK
SMITH
NEALE'S DAM
COVELL MTN.
ROUND POND BK.
MUD POND
TOTE RD.
HEDGEHOG NUBBLE
PROSPECT MTN.
METALLAK LODGE
PERRY STREAM
PERRY BK.
HOTEL
FIRST CONNECTICUT LAKE
TERRY DAM
RIVER
TOTE RD.
OAK RIDGE
CONNECTICUT
CEDAR STR.
BOG BR.
SOUTH MTN.
TROMLEY HILL
WEST BR.
MIDDLE BRANCH
ROUND TOP MTN.
CONNECTICUT LAKES REGION
SCALE — ONE MILE
LOGGING CAMP
TOTE ROAD
LOGGING DAM

The ice has gone out, the channel has cleared, and now comes the chore of rolling in the remaining logs. Paul Stream in April 1914. Stone and Webster report, 1914.

"Breaking out" a rollway required good coordination by the peavey men.

As the logging came to a seasonal end and the men began to drift out of the woods camps, the ice was still held tight on the lakes and streams. This opportunity allowed the idle logger to have a prolonged fling and spend much of the hard-earned money paid him at season's end. Or maybe he headed instead for the boarding house up river, but regardless, a close eye was kept on the river. Then, as the headwaters of the Connecticut River began to swell from the melting snow, the current could be seen forcing its way up through openings in the weakening frozen river surface. Moderating air temperature was softening the ice surface and pressure increased against the bottom of the river ice. Calls came down from the logging camps that the ice on the streams was breaking and the banks were full. The bosses of the drive gathered the whitewater men, hustling them out of the boarding houses, and saloons, sobering them up. Get over to Connecticut Lakes without delay and check out your tools, they demanded.

Then it all happened, too suddenly. Thrust loose by the swelling spring current the river ice rapidly dislodged, sometimes in large chunks that would threaten anything in the way. It was then that the large crew of assembled drivers had to move fast; there would be no days of rest now for the next three months or so. The rows upon rows of logs that had been piled on the frozen surface of the river and streams began to peal into the current on their own, row by row, as the rising water gurgled and roared below and around the piles. The ice had first let go over the current and then had worked loose toward the banks, releasing more logs into the bubbling water.

However, for the masses of logs piled higher up upon the banks it meant brute force on the part of the men to get them into the current. And they had best get them in there before the spring runoff subsided and the water level dropped. The logs in the piles would usually begin to roll into the river on their own once a couple of men had pried loose a few logs in the front of the pile with their peaveys. But there were always a number of logs that had

Both peaveys and pike poles were used to push the logs out into the current.

to be laboriously rolled into the river current. And the boss would allow none to be left behind.

So there was no rest now. Not until all of the logs were into the Connecticut and on their way down the river. The ice had already gone out of the lower part of the river, of course, possibly many weeks prior to the break-up on the upper Connecticut. But this main part of the drive had to wait for the open water on the north end of the river.

As mentioned, the softwood timber in the North Country was almost entirely spruce and balsam fir, and the river drivers liked it that way. Spruce logs float with one-quarter to one-third of the log above the water surface; pine logs have a higher specific gravity and float almost submerged, especially if still green. Thus pine logs require deeper water for flotation, which was one of the reasons why the rivermen preferred spruce.

THE RIVERMEN

The Connecticut River log drive would start out with a small army of dedicated men working down along both banks and out on the river itself. The drive would frequently commence with as many as 500 to 700 men, although that number would be pared down to less than 200 by the end of the drive when less manpower was needed.

River drivers were tough, bronzed and hardened to the privations of three or four months of working in icy water up to their waist for as many as 16 hours a day, eating while sitting on a wet log, and sleeping on the river bank. The men were rough in manner and speech, but they loved the hard work of a river drive.

Much has been written about the river driver's agility and fearlessness, frequently written with the writer's license for exaggeration to emphasize the skills of the men. Romantic historian Stewart Holbrook, for example, wrote that you could throw a bar of yellow soap in the river and the river driver would ride the bubbles to shore. But their agility and skills were renowned; true it was that the ability to swim might be of little value if a man fell in among the fast moving logs.

The riverman was strong but not musclebound. To survive he had to sharpen the skills of balance, agility, judgment, quickness and endurance. A good riverman developed what was called "log sense." Through experience, he could judge the tendencies of currents, water volumes, locations where logs might jam and how to avoid a jam, how to break out a rollway of logs without being crushed, etc.

No matter what task he was doing on the river, the log driver had

Quick feet were needed to stay out of the river, also alertness because this was a bad place to be if the logs suddenly started moving. Collection of A. Janion.

to be alert at all times. There was no time for relaxation. Riverman George Anderson from Houlton, Maine, experienced this in a tragic way. In 1915 he was riding a log through a fast current at Perry Falls, a short distance below the dam at First Connecticut Lake, and his relaxation turned out to be a fatal error. Letting his peavey drop down too far, it hit the current and caused him to lose his balance, throwing him into the river. Anderson grabbed a log that came along but made the mistake of trying to hoist himself up onto the log which then kept rolling out of his hands. If he had just held onto the log end he might have been able to float down to where men were waiting to help him. But the cold and shock overcame him as he struggled, and the current swept him into and under a log jam for a tragic end.

It has often been said that a riverman rarely caught pneumonia, such was his conditioning and physical stamina. That may have been so, but it was a different story with rheumatism, which afflicted many in later years.

A few rivermen were sent to "tend out" at river bends where they would pole logs away from the "off" bank, lest they halt in an eddy and create a jam. The man in the middle is straightening out a "gill poke" in the center of the narrow current to prevent a bunching together. Collection of A. Janion.

For clothing, the rivermen would wear heavy pants, usually kersey or corduroy, torn off or "stagged" above the boot for ease in running along the logs, and with a plug of tobacco in the hip pocket. Red flannel shirts, brass studded belts and broad rim felt hats were also common.

The laced boots were fitted with 50 to 60 small spikes or "caulks," from ¼ to ¾ inch in length, on the bottom of boot for gaining a secure foothold on the logs they walked upon. They wore these boots all the time and the evidence of the caulks was seen everywhere in the towns along the river, with splinters on the wooden sidewalks and pockmarks on the floors of the stores and saloons.

An unidentified North Country logger in typical dress.

Continuous immersion in the water would cause the log driver's feet to parboil and turn scaly. Hoping to find some relief, the men would put white lead in their socks or make use of the tub of lard placed near the tent in the morning. Riverman Bill Dexter, who started work on the river at age 17, said he wore sneakers most of the time.

There was an interesting mix of ethnic groups among the men going down the river. In the early years of the drive the work force was predominantly composed of the "Bangor Tigers," a hodgepodge of experienced rivermen from Maine. During the 1870s and 1880s there was a large influx of French Canadians, who were outstanding axemen but frequently not as competent with horses. The migration from Quebec let up in the 1890s but picked up again in the early 1900s.

Through the years there was always an influx of men from Maine, New Brunswick and Prince Edward Island (the PI's), lumbermen that couldn't resist another season on the big river drive. Many were Irish, well known as excellent rivermen and loggers who proved to be tough and fearless. From New Brunswick there were occasionally a few blacks, also said to be good rivermen.

A visitor to a logging camp near Second Connecticut Lake in 1895 was impressed with the ethnic mix among the loggers. As identified in his photo, the men seated are (1) American, (2) Italian, (3) Irishman, and (4) Frenchman. The small man seated on the far right was the cook's wood chopper who didn't want his "picter tooken" and thought he was out of the range of the lens.

Perley Hurd remembered one named Coxee, who always sat, ate and camped alone.

Regardless of where they traveled from, the rivermen had to be there when the ice went out. A small gang had already been sent ahead for preparatory work down river. There were always booms to build, sorting gaps to prepare, and food purchases to arrange, all to be accomplished before the main body of logs and men appeared.

Assembled at the head end of the drive with the hundreds of men were scores of horses and wagons, tons of provisions to move along the river course, great supplies of axes, peaveys, pike poles, and tents. The men were divided up into two sections, the larger group to travel down with the main body of logs and a smaller group to wait and work along the rear, to clean up. The men in the main body were divided up into groups of about 50, each working under a boss.

The overall drive boss had a big responsibility; unforeseen circumstances would put him to the test, and success or failure might well depend upon him. The drive bosses became legendary with their deeds of skill and daring and were usually to be found at the rear of the drive where most of the men were working.

There was John Stapleton, from Veazie, Maine, who began work on the Connecticut in 1875 at age 22, and Sam Watts, who worked on the river for 19 years and was first selected as drive boss in 1893 after serving previously as head clerk. Fred Gilmore started work on the river in 1862, the same time as Abner Bean and John Ross. Not to be overshadowed was the storied Winfield "Grinner" Schoppe (1850–1924). Schoppe started on the Connecticut in

Tools of the Rivermen

Peavey: Made of a long wooden handle, often six feet long, with a metal point fastened on the end and a swinging cant dog hook for rolling logs. Developed in 1870 by a blacksmith in Maine named Peavey. In 1887 the Peavey Company in Bangor, Maine, shipped 2,500 peaveys to the Connecticut River Lumber Co.

Pike Pole: A long wooden handle with a long metal spike on the end. Handle length was usually about 8 feet but could be as much as 12 to 20 feet. The metal tip had a slight twist to it, so that when driven into a log the point would hold enough to permit pulling of the log as well as pushing.

Marking Axe or Hammer: Used to brand or mark a log on the end with the identifying mark of the owner or company.

Log Auger: Holes were drilled through the log near the end for attaching boom chains or for pinning logs together in a raft.

Artist's rendition of a log auger.

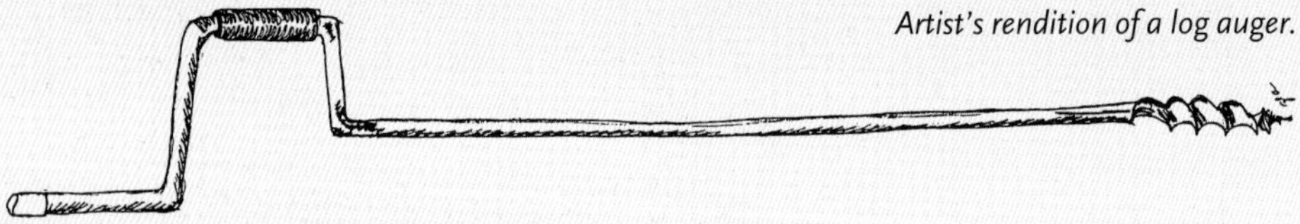

Artist's rendering of a Peavey.

Every "river hog" had to know how to bundle the long-handled peavey. This photo is from the 1912 log drive. Collection of A. Janion.

Pike poles were essential for pushing and guiding the logs. The men on the catwalks are urging logs through the sluiceway in the dam.

Artist's rendering of a pike pole.

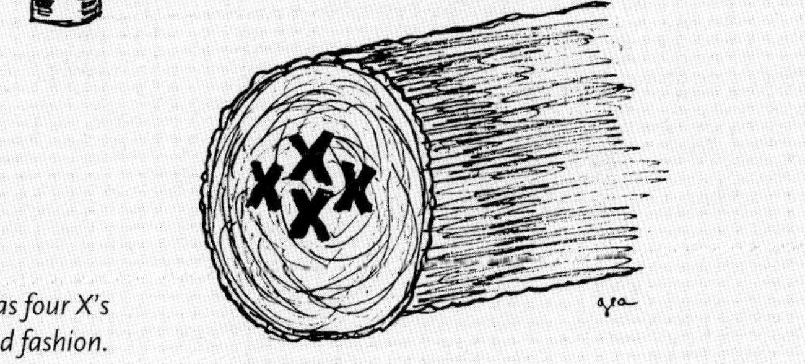

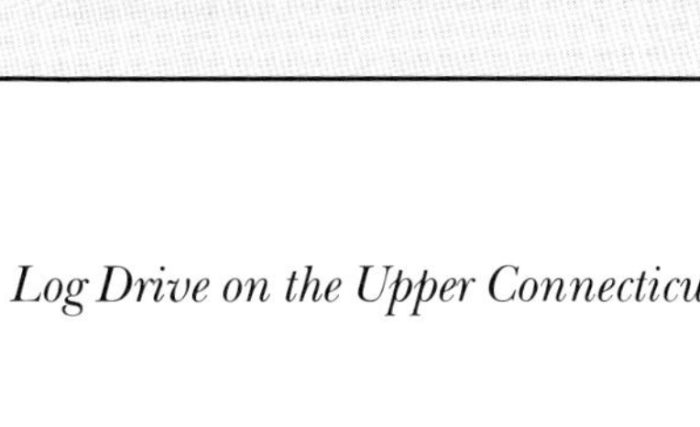

The CVL Co. log brand was four X's placed in diamond fashion.

Many of the old-timers vowed that the quickest riverman on two feet was the renowned Dan Bosse, shown here at age 57.

1874 and became renowned as a dam builder, among other accomplishments. There were also Frank and Bill Smart, "Big" Phonse Roby, a respected "walking boss" and master driver, and Tom Graham. George Van Dyke's brother, Philo, also served some years as a drive boss. Among the foreman working under the drive boss one would frequently find Charlie Leighton, Charlie Robie, Joe Roby, Jim Doherty, and Jigger Johnson.

The beginning or head end of the log drive was the Connecticut Lakes in the northern tip of New Hampshire, although in later years the main drive began a ways further down the river as the timber was cut off up above. During the early years of the river drive era the Connecticut River Lumber Co. was cutting heavily around the First and Second Connecticut Lakes and sent large quantities of logs over the dams at the foot of the lake to begin the drive.

Booms of logs would be pulled across Second Lake, sluiced over the dam, floated down the stream into First Lake, and boomed again for the haul down to the large dam at the outlet. The three-mile-long stream between the lakes had lots of hang ups, so when the logs were released from Second Lake there were men stationed all along the bank within sight of each other. When a jam occurred the message was relayed by hand signal to close the gate at Second Lake.

In the 1890s, when the company was logging along the East Inlet to Second Lake, a dam needed to be built on the East Inlet

Jigger Johnson was renowned for his toughness in the north woods and in the river drives. Many highly exaggerated tales of his exploits have evolved over the years.

about two miles above the lake. But the area was mired in quicksand, and piling needed to be driven in to stabilize the dam. The problem facing the company was figuring out how to get piles driven on site when the nearest railroad station, where a pile driver could be delivered, was 56 miles away.

First off, a good team of horses was sent down to the North Stratford station and in five days was back with the large pile driver. To get the machine across the lake, two rafts of logs were specially built to haul the machine, the horses, provisions and a crew. The

Win Schoppe is seen here on the far right along with other river bosses and guests at a cook tent circa 1900.

The dam at Third Connecticut Lake as it appeared in the early 1920s. This was the uppermost dam on the river, at an elevation of 2,191 feet.

The outlet dam at First Connecticut Lake was built in 1880–1885 of sawn hardwood, principally birch. There were three 18-foot gates in the 410-foot-long dam and two 13-foot-wide spillways on the south end with an inclined plank deck over the spills. The storage height of the water was 18 to 19 feet. Stone and Webster report, 1914.

Booming Logs Across the Lake

Kedging or moving a mass of floating logs across a lake to the outlet required a special technique. About two to five thousand board feet of logs were herded together and enclosed within a boom or string of large logs that were fastened together end to end. The fasteners in the early days were thoroughshot pins, later replaced by chains.

The boom of logs was moved across the lake by a headworks located on a raft made of hewn timbers piled crossways in layers. The raft was moved out beyond the boom and secured by an anchor weighing 200 to 300 pounds. Then the boom was pulled toward the raft by a primitive capstan mounted on the raft that would spool in the long warp attached to the boom. The spool was a short hardwood log set upright on a shaft with capstan bars set horizontally into the top. A large chunk of salt pork at the bottom acted as a lubricant.

Rivermen at a camp on Second Connecticut Lake in 1895 display an anchor used for moving a boom across the lake.

When spooling up the warp, two men would work on each bar as they tediously walked in a circle, occasionally even falling half asleep. They had to remember to step high each time they came around to where the hawser came up to the spool.

When the boom was brought up to the headworks, the raft would be moved out again to a further location, about a dozen men in a boat lifting the heavy anchor and hawser to move it and the raft along the lake. In this manner, step-by-step, the large mass of logs was moved across the still waters. But it was difficult work, especially against a wind, which would sometimes blow the whole mass of logs back to the starting point.

Towing a boom on First Connecticut Lake with a motorboat.

Upon reaching the foot of the lake the boom was opened and the logs were released to go over the dam and down the river.

In later years the Connecticut Valley Lumber Co. had a motorboat on the Connecticut Lakes to pull the log booms.

Roll dams were constructed on the stream between Second and First Lake to back up water and flood over obstructions. The logs would roll over the top. Stone and Webster report, 1914.

rafts were hauled by a bateau powered by eight rugged oarsmen, with George Van Dyke steering. The mile-and-a-half trip across the lake to the far shore was covered in two hours, and from there the horses completed the remainder of the haul.

The new dam stood for many years, but logs were cut and driven down the East Inlet only once, in 1898. The low water that spring and the lack of other upstream water storage caused many of the logs in that drive to be hung up along the banks of the Inlet between the dam and Second Lake. Some of those logs could still be seen there in the 1940s.

Below the dam at First Connecticut Lake there weren't many troublesome spots, except for Perry Falls, but there was always danger lurking nearby for the unwary or careless one. Rapids or falls or any other obstruction that would alter the currents were always locations for caution.

Bridge abutments in the river were definitely troublesome objects for the river driver; logs would collect against them and create a jam behind. Log masses have been known to take out solid concrete abutments, even moving an entire bridge along down the river. For that reason, it's not difficult to see why George Van Dyke and the CVL Co. were often at odds with the railroad companies and the town selectmen, and why a number of lawsuits involving the CVL are on record.

The railroad bridge across the river at North Stratford once backed up a logjam that took Phonse Roby four days to unravel, even though he had hundreds of men on location. For a time the railroad company was worried they would not be able to run trains across the span all summer long.

Rocks and ledges along the river were also troublesome in times of low water flow. August of 1879 found a gang of 27 men camped at the river headwaters blasting out obstructions. The next year they were at it again, blasting rocks at Healy Falls and Lyman Falls in Bloomfield, Vermont, as well as at Beecher Falls.

Constant Troubles with the Railroads

- In April 1894 a log jam caused the Connecticut River to overflow and undermine some Grand Trunk Railway track east of the North Stratford, New Hampshire, railroad station. The next morning, Sunday, a work train with 20 flatcars and 30 men was on the way over from Island Pond, Vermont, to make some quick repairs. At 6:15 A.M., while passing over the Nulhegan River, the work train fell through the iron bridge; rushing water had weakened a stone abutment. The engine and 14 cars tumbled into the chasm, drowning brakeman Thomas White.

- On April 20, 1914, a CVL log jam forced water up onto the embankment along the Grand Trunk line near Baldwin's Cut, near Beattie Station, causing a large washout. The river had risen quickly after a warm spell.

- The U.S. District Court had to assess damages in a 1913 case that involved a river overflow with logs floating onto the track and washing out 3,000 feet of the roadbed. Damage was set at $25,000.

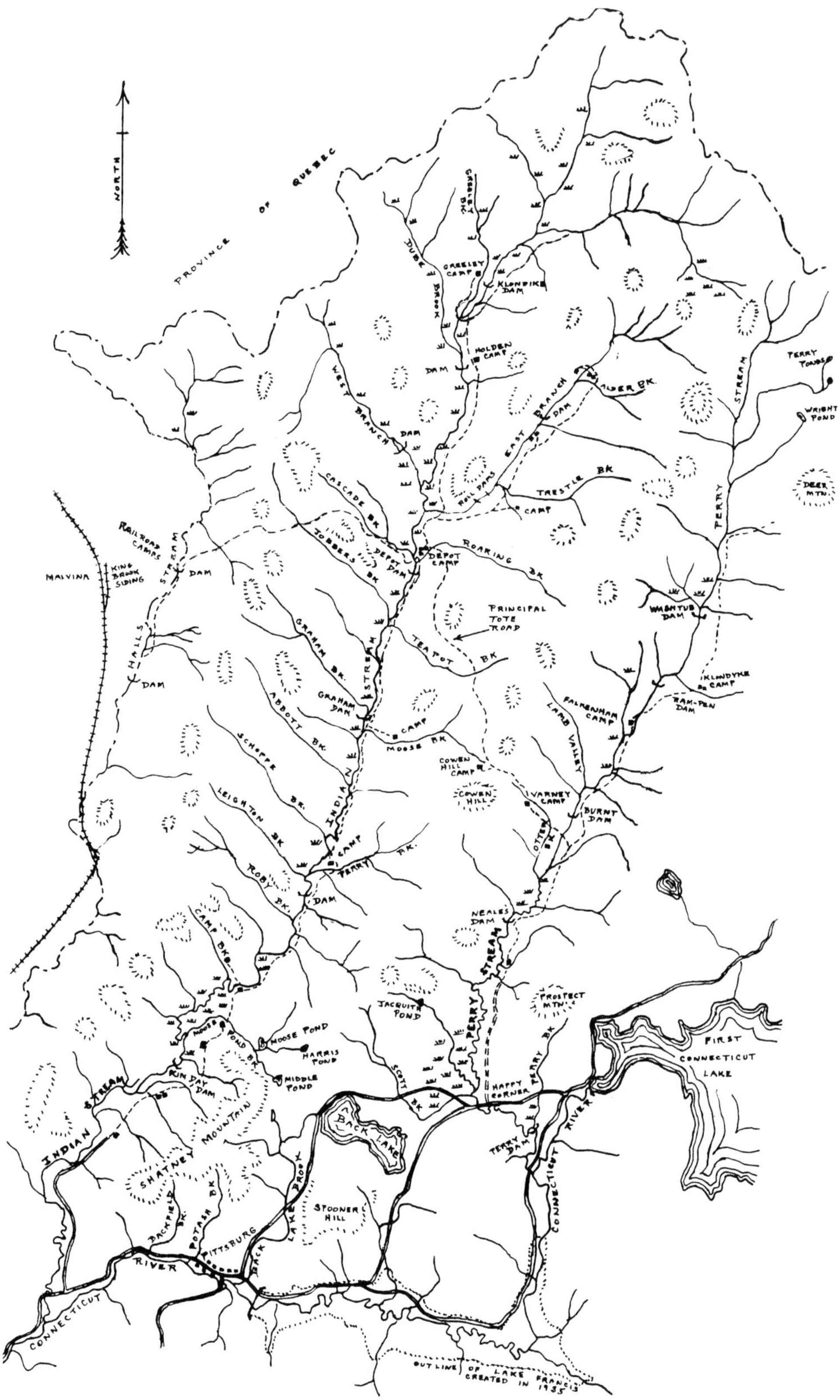

INDIAN STREAM AND PERRY STREAM

- LOGGING CAMPS
- TOTE ROADS
- LOGGING DAMS

SCALE
ONE MILE

Residents of Canaan, Vermont, watch apprehensively as a jam of logs and ice enlarges against the covered bridge across the Connecticut River in 1900.

MEALS ON THE RIVER

Feeding 600 or so hard working men four meals a day and doing it on the move down the river was not a simple task. The food had to be there, and the men accepted no excuses. Four meals a day were standard: breakfast was served at about 5 A.M., first lunch at 9 or 10, second lunch at 2 or 3 P.M., and supper about 6 or 7 P.M. And since all the men could not be at the same place at the same time there could be as many as six separate meals served daily. The river, of course, wouldn't stop flowing on Sundays, thus there were no days off.

A portion of the Canaan bridge, as seen on the left, had formerly been taken out by a log jam. The toll house is behind that section.

The wanigan with the cook tents as set up along the road in Guildhall, Vermont.

The meals served the rivermen were sumptuous, with baked beans, fresh meat, milk, bread and other baked items. All were prepared despite the constant need to move the tents and cooking facilities downriver, often on a daily basis. Baking was accomplished with an open front Dutch Oven set before the open fire and adjusted to different angles as needed. There were large cook-

Bean Hole Baked Beans

If the cook was allowed a few hours in one location he would often prepare a mess of baked beans cooked in a bean hole. A large hole about four feet across would be dug while the local boys gathered around watching with fascination. A brisk fire was built within, and when hot, a number of stones were placed within the fire. As the fire slowed down and the stones became red hot, a black iron kettle containing parboiled beans was set on the stones with a lid secured tightly. The beans were likely seasoned with salt pork and black strap molasses. The kettle was then covered over completely with ashes and dirt and left for a few hours. The cook knew from experience just when to uncover what was reputed to be a feast of baked beans unparalleled in delicious taste.

Chow time at the wanigan. Note the bean hole in the foreground and the reflector ovens behind it.

ies, hot gingerbread with maybe a few cinders on top, and if the camp was stationary for a time, pies and doughnuts too. The CVL's floating wanigan, dubbed "Mary Ann," had a large kitchen range set up within the cabin for use along the lower part of the river where a barge could be floated.

The cook's tent is set up on an island in Brattleboro, Vermont, during the 1902 drive. A log jam is noticeable in the background.

Many supplies, such as milk, eggs, fresh beef and potatoes, were purchased from local farmers as the drive made its way down the river. Not unusual was a daily consumption of 60 dozen eggs, a bushel of beans, or a bushel of doughnuts, with food expenses running as high as $2,000 day. The responsibility of making sure enough food was on hand fell upon the manager's clerk. Charles Hadley, the CVL clerk in 1912, once received the unwelcome report from cook Jim Belanger that there was no milk on hand. Since the crew would consume as much as a barrel of crackers and 120 quarts of milk at the evening meal, Hadley had no choice but to hire an automobile for a quick trip from the camp location just below Putney, Vermont, to nearby Brattleboro for supplies.

Obtaining the supplies took well-planned organization. Storekeepers such as John Pattee of North Stratford would order and stock the needed supplies. And Pattee would send a man down the river ahead of the wanigan (supply wagons) to buy fresh milk, eggs and other essentials for the next camp location. Cook Wilbur Mills

The cook for the International Paper Co. drive on the White River provided freshly baked bread.

would use a barrel of flour a day while cooking on the floating Mary Ann.

The men would have to eat at a convenient location wherever they were working along the river, and lunches were usually brought to them by the cook's helpers. If it was raining the men often ate standing up with the rain running off the clothes into the food. The cook would stir the kettle of meat with rain dripping off his dirty hat and down into the contents of the kettle.

The cooks and cookees were not particularly agile themselves as rivermen. In one instance, Charles Buzzell of St. Vital, Quebec, a

The rivermen enjoy a quick meal on the river bank.

twenty-year-old cookee, drowned on June 16, 1887, at Brattleboro while attempting to cross the river on the logs.

There was no lingering at the early morning breakfast meal in the pre-dawn half-light, with the cold mist hanging over the river. Each man would pick up a tin plate and a tin pannikin for tea and pass by the cook, who had his kettles hung over a fire and tin bakers on each side filled with biscuits. If the men delayed with the meal, legendary drive boss Tom Graham might holler "that cool ice water won't hurt ye. Just think, when the drive is over you'll be paying good money to a bartender to serve it to you with a little whiskey in it."

A knoll above the Wilder dam was an ideal location for the river crew, which would spend a few days here at the important Wilder Boom. The log drive is being temporarily held behind the boom.

As the water level in the river fell, the crew had the job of picking up the stragglers at Guildhall Dam in 1910. The men on top of the apron are about to pass down a couple of boats. Courtesy of Lester Fogg.

The logs continued their journey, day and night, in a steady flow. As the logs passed the upriver villages of Colebrook, North Stratford and Lancaster there were few problem locations. The situation was soon to change, however.

As a youngster, author and historian Stewart H. Holbrook lived along the river in North Stratford and well remembered the excitement in town when the drive was about to go by. There was the great booming sounds that could be heard in town during the night, and the excitement of a youngster eagerly awaiting daybreak as the drumming sound of logs pounding together in the churning water became louder and louder. As the drive drew nearer, every boy dashed to the riverbank to watch their heroes in caulked boots and red shirts. They dreamed not of being a cowboy or a policeman or the engineer of a speeding steam locomotive; their ambition was only to be brave rivermen and travel on down to places wherever the river flowed.

Even before the log drive actually reached the various towns along the river, there were always early indications that the arrival of the spring drive was imminent. An occasional log, an "early traveler," might appear as it had somehow jumped ahead of the pack, the supply wagons could be seen along the town roads bordering the river, or the boss of the drive, possibly even George Van Dyke himself, would appear proudly riding through the village. Then came the logs relentlessly pushing on, miles of them filling the river surface from bank to bank. An advance contingent of rivermen would be stationed at troublesome river bends where the logs might jam, or in an eddy where the logs could lose momentum and accumulate. In either location, the men would be found poling the logs out into the current with a pike pole.

Should the drive get held up behind the main crew of rivermen, however, pity the village folks. One of Van Dyke's timekeepers once recalled how a crew of almost 300 men were in North Stratford one night, awaiting the next section of the drive. As he described the scene, "You could stand on a piazza roof and look down into the square, packed with 500 men, and count fifteen fights going on at once." When the logs came, however, the men were all business again.

FIFTEEN MILE FALLS

For the first part of the drive below the Connecticut Lakes there were not many troublesome areas along the river, the exception being the occasional bridge abutment. It wasn't until the drive reached a location about ten miles below Lancaster, usually in the early part of June, that the troubles began. Here, the men encountered what was probably the most treacherous stretch along the entire river journey. A succession of rapids began just below Fitzdale (Gilman), Vermont, across from Dalton, New Hampshire, and the foaming whitewater stretched for some 20 miles down to East Barnet, Vermont. The river elevation fell 370 feet in this dreaded stretch, along which there were many rocks and ledges for the logs to hang up on.

Throughout this stretch the rapid current kept the logs in constant conflict for space as they careened, jostled and smashed into each other and the many boulders in mid-stream. The impacts were enough to occasionally hurtle a log right up into the air. Logs caught for a few seconds between the boulders might quickly back

A stretch of Fifteen Mile Falls as it appeared during low water in August 1928. This site is the location where Comerford Dam was built in 1930, backing up eight miles of deep water.

up hundreds of other logs. They called this section of the Connecticut River the Fifteen Mile Falls, a name that appeared on maps even back in the mid-1700s, despite the fact the stretch was at least twenty miles long.

The falls were considered so treacherous that boats were not taken down through them. Instead, the rivermen arranged to have their boats carried around the falls on horse-drawn wagons. The company would also have a boom strung across the river at the head of Fifteen Mile Falls, which was used to shut off the flow of logs in the event of a jam down below. Eighteenth-century canal builder DeWitt Clinton once favored the construction of a canal around Fifteen Mile Falls, or so it has been written.

As the drive reached Fifteen Mile Falls, the seasoned veterans

Fitzdale

Fitzdale was the site of a paper mill village created in 1898 with the construction of a wooden dam and a pulp mill built by the Dalton Power Company. The village later took the name of Gilman when Russian immigrant Isaac Gilman became a stockholder in 1910, and the company president soon afterwards. In 1913 the name was changed from Parks Paper Company to Fitzdale Paper Company and in 1921 to Gilman Paper Company.

In the early years the paper mill received most of its softwood supply by way of river drives on the Connecticut River. In the 1940s rough pulpwood was arriving by truck and was run through the company's drum debarker at the pulp mill. Each spring the company would call the St. Regis Company office at West Stewartstown, New Hampshire, to assure that any pulpwood drives on the Connecticut River had been completed. Then the company would dump their wood into the river to soak the bark before running it through the debarker.

The Gilman Paper Co. mill and dam at a later date.

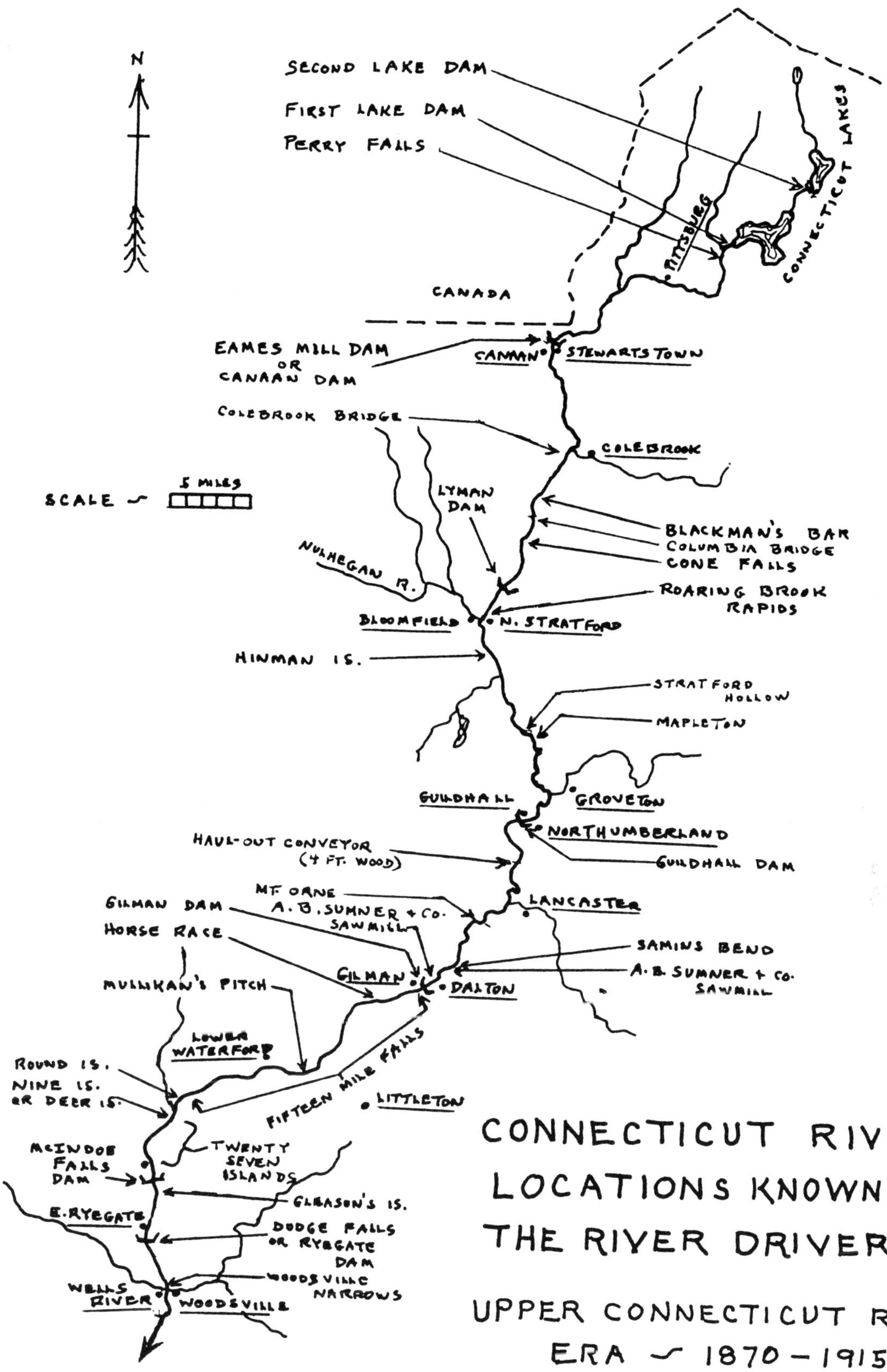

CONNECTICUT RIVER LOCATIONS KNOWN TO THE RIVER DRIVERS

UPPER CONNECTICUT RIVER ERA ~ 1870 – 1915

CONNECTICUT LAKES TO WOODSVILLE

The bridge between Dalton, New Hampshire, and South Lunenberg, Vermont, could be troublesome. Here, a crew is attempting to loosen the jammed logs with dynamite.

would carefully work their way down alongside the rapids, releasing jammed logs with the ease and confidence gained from years on the river. Prying free the key log in a jam—the one holding back the other logs—was the hardest chore, because once released the entire mass could quickly move along with the current, carrying along the men who rode on them as they simultaneously ran and jumped over the tossing logs to make their way to shore.

Fifteen Mile Falls was certainly no place for the rookie riverman. An unsubstantiated claim has been made that 13 lives were lost one year along on this stretch, either on jams or in the treacherous current. Another unconfirmed report has 20 lives being lost in a single year. Those claims can't be verified, but there's no question but what those rapids took many lives. On the other hand, in 1891 a drive of 70 million board feet went through the Falls, tended by 400 men, and not a life was lost.

The log drive of 1882 was a particularly difficult drive because of high water in the river; according to one newspaper report, by the end of June eleven lives had already been lost since the start of the drive. That year the drive boss reportedly had 150 men working at Fifteen Mile Falls and 250 toiling just above.

Legend has it that when the bodies of drowning victims were recovered, they were stuffed into a pork barrel and buried on location. Historian Robert Pike claimed that in 1930, when New England Power Company was doing excavation work for its new hydroelectric dam at the foot of Fifteen Mile Falls, the remains of some barrels with spiked boots still inside were discovered.

The worst of the worst along Fifteen Mile Falls was a quarter-mile long succession of fast water among boulders, near the beginning, known as the "Horse Race." It was what Stewart Holbrook called the "milltail of hell." Legends abound and grow about brave feats when it comes to the dangerous deeds such as riding a log through the Horse Race, but one story has it that Bill Bacon was the only man who did so and came out alive. Another legend has riverman Skiff Anderson as the sole man with such an honor.

Rivermen, of course, were always susceptible to a challenge. Back in 1869, Warren Hilliard of West Stewartstown was taking a drive down to the mill at McIndoe Falls when he and George Christie of Canaan attempted to run a bateau down through the Falls. Both men drowned and Christie's body was found ten miles down the river at Wells River.

A little more than halfway through Fifteen Mile Falls there was a bridge across the river at Lower Waterford, Vermont. On May 7, 1890, the logs jammed against the bridge abutments and eventually collapsed that half of the bridge on the Vermont side. Twelve hours later the other half went down. Another bridge at Upper Waterford was replaced with a single-span steel bridge that was later taken out in the great flood of November 1927.

Riverman Fleetwood Pride also wrote about a jam at the Waterford bridge in his memoirs. The boss of the drive that year was Fred Gilmore, considered by Van Dyke to be one of the best drivers in the country. Gilmore was on hand to personally direct the breaking of this jam and was using horses to pull out the logs that were wedged in tightly at places deemed too dangerous for the men to work. A long rope with a hook on the end, called a "dogwarp," would be attached to the log by winding it several times around the log and driving the hook lightly into it. As the horses on shore would pull the rope, rolling the log free into running water, the log would keep rolling over until the rope came free of the log.

Gilmore, however, suffered a humiliating injury on this occasion. He was standing near the straining rope, cursing the men for being afraid of the line, when the rope splice at the end by the hook let go, allowing the tensioned line to snap back against his face. The blow ruined his right eye and permanently damaged his left one, prematurely ending his career on the river.

What loyalty Gilmore may have had to the company was not enough to keep him from suing Van Dyke for providing rope that the company knew was unsafe. Gilmore brought a lawsuit against CVL in 1886 and was eventually awarded $20,000 in damages.

At the end of Fifteen Mile Falls, where the Passumpsic River flows into the Connecticut, there was another difficult stretch of river known as Mulliken's Pitch, and below that, the Twenty-Seven Islands, another series of hindrances that taxed the skill and hard work of the rivermen. Legend has it that these locations were responsible for 20 deaths over the years, while Perley Hurd rated Mulliken's the most dangerous place on the Connecticut.

Fifteen Mile Falls does not exist any longer. As the New England Power Association continued to harness the power of the Connecticut River for production of electricity, the rapid drop of Fifteen Mile Falls offered an excellent location for power development and was the subject of engineering studies for many years before being developed.

At the foot of Fifteen Mile Falls a 170-foot-high dam was built in 1928–30, known as Comerford Dam, which ponded the river back for eight miles. Then in 1957 a second power station (Moore Dam) was completed about midway through the former Fifteen Mile Falls, ponding the river for another 12 miles above and completely inundating the treacherous rapids that had claimed so many lives over the years.

LOG JAMS

Log jams could occur just about any place there was an obstruction in the river, and once a few logs had their forward motion stopped as they came against a boulder, ledge, or bridge abutment it didn't take long for a mass of logs to bunch up behind them. A strong current would roll the logs and pile them up on end, jamming them together many logs deep. The fir logs would tend to sink sooner and wedge crosswise in the current.

Then the hard and sometimes dangerous work began, for the men had to go out onto the face of the jam and pick it apart. With peaveys and pike poles the restraining logs had to be pried out; quite often, too, there were only one or two key logs that had to be tugged out. In freeing up the jam, as many as 20 men might have to drive their pike poles into the one key log to yank it out together, or they might have to move as many as 50 or more logs to clear the jam and get the logs moving again.

Some jams, like the one shown here, had to be picked apart one log at a time.

The front end of a square jam would have logs piled high and pushed well up into the air. The river water could be backed up into the villages above.

Once the front of a jam was unraveled the entire mass would likely begin to move along, sometimes quite suddenly and forcefully. With nimble feet the men would leap from log to log in a mad scramble for shore, lest they be carried into rapids down below. If there was relatively calm water below they might even choose to ride along on the slowly moving log mass.

Breaking up logjams was a hazardous job, but one of the many expected duties of a riverman, and over the years many lives were taken as a result. As a logjam broke, a man could easily find him-

"Cutting a hole" in a log jam required a united effort of many hands. Stone and Webster report, 1914.

Types of Log Jams

The river drivers would classify a logjam by its severity and the location in the river.

SQUARE JAM: Logs were tightly wedged together entirely across the river. The jam would be thickest in depth at the obstruction, reaching from the river bottom up to six feet or more above the water and then taper down toward the upstream end.

CENTER JAM: A stoppage in the center of the river such as a ledge that would create a back-up but leave a free channel on either side.

WING JAM: A stoppage against a hindrance near the shore, such as a ledge or shallow area, leaving a portion of the river open. Wing jams were frequently allowed to remain in place until the final crew came along cleaning up the rear of the drive. A wing jam would deflect more water into the main channel of the river, deepening the flow over shallow areas. An entire series of wing jams might be encouraged to form and then left be in place if there were a lot of ripples in the main channel.

DRY WING JAM: A jam that formed during high water but was left on dry ground when the river waters receded. This was the most difficult jam to break up because the logs had to be dragged back to the water.

The river appears to be completely blocked by this massive log jam. It looks like the crew is preparing a dynamite charge.

Both a center jam and a wing jam are evident at Sumner Falls in Hartland in 1912.
Collection of A. Janion.

Four Causes of Log Jams

1. A narrow location in the river with fast (quick) pitches of water, frequently strewn with boulders or shallow to gravel, would obviously hang up the logs. On smaller streams this could be alleviated by constructing a roll dam immediately below the quick stretch in order to back up water and flood out the shallow spots. Low water level in the river also resulted in more frequent jams.

2. Any obstructions in the river such as a ledge or boulder were particularly troublesome when there were many logs in the river. On small streams it was common for the company to blast out the boulders.

3. Bridge abutments placed in the main path of log flow were always troublesome. Abutments were often built with the upstream face in a wedge or pointed shape, called a "cutwater," to deflect logs and large chunks of ice. But they weren't always successful in preventing jams.

Logs become lodged at the piers on the railroad bridge between White River Junction, Vermont, and West Lebanon, New Hampshire.

4. A sudden freshet upon a river bed already filled with logs from bank to bank, and too great to be controlled by upstream dams, would cause the upstream piles to work against those immediately below and jam them tightly together.

5. When the center channel in a stream was free of ice before the downstream stretch, the first logs could lodge against the ice obstruction below and form the nucleus of a jam. Then, if the water rose suddenly, many logs might come down "en masse" to add to the jam. The problem was usually corrected by blasting out the downriver ice.

Boulders or ledges in the center of the river caused frequent log jams.

self caught on a small group of moving logs with no way to shore and pure disaster ahead. Al Patrick, drive boss for many years, said he had seen as many as 13 men killed on one drive. But as he stated, "It was almost always carelessness to blame and it don't do a logger much good to know how to swim." A local newspaper also reported that in a single tragic accident at a jam in 1914, six men died.

The drive boss would be careful not to send an inexperienced man out onto the jam for the lives of other drivers were often at stake. Phonse Roby was once watching one of his men working apart a small jam when the logs suddenly began to move in the current. Instead of scrambling for shore, the man knelt down on the moving logs and stayed with them. The logs soon rolled over and dumped the rider into the current. But the water was deep enough for the logs to pass over him, doing no damage except tearing his clothes. Roby was not impressed; he told the man to collect his time and leave the drive immediately.

Jammed logs were obviously no place for novices either. Local folks once enjoyed a good chuckle over the experience of Harry Hartshorn, the railroad station agent in Lunenburg, Vermont. While on his way to church one Sunday in the spring of 1878, he observed the rivermen working to break up a logjam. He foolishly walked out on the logs just before the jam began to break. Scrambling over the logs toward shore, Harry got dumped into the river before finally reaching dry ground. As reported in the local gossip column, Harry's pew at church was unoccupied that morning; his best clothes were no longer presentable.

One of the worst Connecticut River jams in history occurred in June 1900 near the village of Wells River, Vermont, when 55 million board feet of logs jammed up against some bridge piers, piled solid for a half-mile up the river, and in some places 6 to 10 feet deep. About 100 men had already been discharged at McIndoe Falls, but there were still 175 men and two donkey engines on hand to work the jam apart.

If a jam could not be safely broken by his men, the drive boss would usually order it to be blasted apart. Using dynamite was usually a last resort for the boss because some good timber would be damaged by the blast. By the same token, lives could be saved by blasting apart a jam.

The dynamite used in breaking up a jam was rend-rock or "rack a rock," 75 percent nitroglycerin, in half-pound charges. In the early years the rivermen might bore a hole in a log to hold the charge, but a method of tying three sticks of dynamite on the end of a long pole in order to work it down inside the jammed logs was later developed and used. The charge had to be placed against the backside of the key log because blasting from the lower side would

Demise of the Littleton Bridge

On May 6, 1870, a jam occurred at the Littleton, New Hampshire, bridge over the Connecticut River when 25,000 board feet of logs piled up against the piers, ponding a large quantity of water behind the logs. The embankment began to wash out and the rivermen quickly built a cribwork dam above to divert the water over to the Vermont side and then floated about 75,000 board feet of logs through between the pier and the westerly abutment.

At first it appeared further problems at the bridge had been averted, but due to improper guidance of the log flow, the jam intensified and eventually took out the bridge. Later, a court awarded damages of $10,000.

not blow it free. The cartridge case and fuse were both waterproof, and the fuse burned about one foot per minute.

"Blowing the jam" with dynamite, as was being done in here during the 1912 drive, was usually the last resort because of resulting damage done to the logs. Collection of A. Janion.

Riverman Dolph LeClair once recalled how he dynamited a large rock out of the middle of the Connecticut River at French King in 1904. To accomplish this task, he built an A-frame against the rock. Inside the A-frame was an entire box of dynamite four feet below the surface of the river. The resulting blast was so powerful that LeClair claimed a large piece of the rock landed a full half-mile away.

As effective as dynamite was in taking care of the jams, the rivermen could only use it if it was not frozen. Thawing out a stick of frozen dynamite was a dangerous chore, for it was brittle and could easily be set off when jarred. Once up on Indian Stream a large three-story house belonging to Kim Day was blown up when he errantly allowed some rivermen to thaw dynamite in his oven.

Getting the blast down behind the key logs was essential for it to be effective. Usually the riverman didn't have to move far away from the blast, as noted in this scene at Dalton, New Hampshire.

George Van Dyke's sawmill crew poses outside of the boarding house.

McIndoe Falls

The small village of McIndoe Falls is located in the Vermont town of Barnet, next to a falls on the Connecticut River that once formed a favorable mill location. The first dam at this site, a wing dam extending from the Vermont shore to a rock ledge out in the river, was built some time prior to 1803. In 1816 the dam was expanded across to the New Hampshire bank.

A grist mill was built on the New Hampshire side, and later, a small sawmill which would raft the sawn lumber and send it down the Connecticut River.

The 450-foot dam at McIndoe Falls was built against a rock ledge in the river. An old grist mill can be noted on the New Hampshire side.

A large sawmill was built on the Vermont end of the dam in 1830, marking the start of many years of sawmilling at McIndoe Falls. It's been said that since this mill was built over the river, it was taxed by the town of Monroe, New Hampshire.

After the original sawmill burned down in 1875, a new steam-powered circular sawmill was built, and included among the new partners of the mill was George Van Dyke. The masterful and commanding Van Dyke soon became the sole owner of what he developed into a thriving lumber manufactory.

By the mid-1880s the McIndoe Falls sawmill had become the largest lumber manufacturer in northern Vermont. About 85 employees were turning out 70,000 board feet of softwood lumber daily, most of it surfaced in the planing mill on site.

In 1886, when Van Dyke became president and one of the owners of the Connecticut River Lumber Co., his assets, including the McIndoe Falls sawmill, were taken into the corporation. In 1902 the corporation was reorganized as the Connecticut Valley Lumber Co. The sawmill was closed down soon after that date, and when the sawmill at Mount Tom, Massachusetts burned down in 1907, the machinery at the McIndoe Falls mill was used for rebuilding the Mount Tom mill. The building at McIndoe Falls was dismantled in 1908.

The 450-foot-long dam across the river was originally made of logs and extended from each shore to a ledge outcropping in the center of the river. By the mid-1880s the old dam was leaking badly and in need of replacement, although it still provided water power for the grist mill on the east end, which operated with five runs of stones. In 1930 New England Power constructed a new power generation dam on this site.

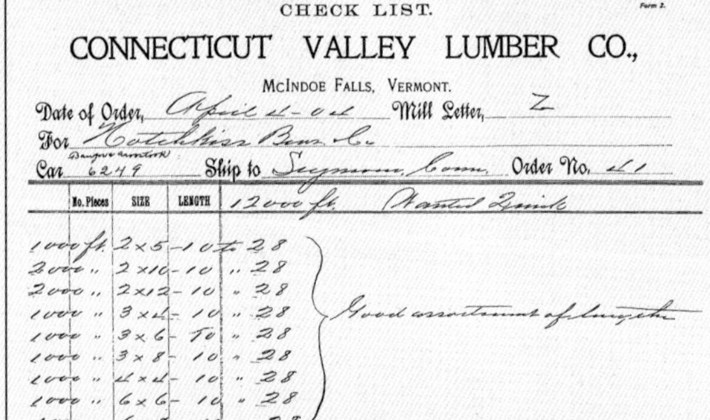

CHECK LIST. Form 2.

CONNECTICUT VALLEY LUMBER CO.,

McINDOE FALLS, VERMONT.

Date of Order, [illegible] Mill Letter, Z

For [illegible]

Car 6249 Ship to [illegible] Order No. [illegible]

No. Pieces	SIZE	LENGTH	
1000 ft.	2x5	10 to 28	
2000 "	2x10	10 " 28	
2000 "	2x12	10 " 28	
1000 "	3x4	10 " 28	Good assortment of lengths
1000 "	3x6	10 " 28	
1000 "	3x8	10 " 28	
1000 "	4x4	10 " 28	
1000 "	6x6	10 " 28	
1000 "	6x8	10 " 28	
1000 "	1x2	[illegible]	

Check List from CVL's McIndoe Falls mill.

Sorting and Sluicing at McIndoe Falls

When the annual log drive reached McIndoe Falls it was usually stopped temporarily and held by a heavy boom of logs stretched across the river at a narrow point about a quarter-mile above the dam. An advance gang of men was sent down each spring to construct this boom.

An opening about 30 feet wide in the center of the boom allowed the logs to pass through into a sorting gap where about 15 to 20 million board feet of logs were sorted out for the company mill at McIndoe Falls. The men doing the sorting stood on a small floating platform or catwalk made of planks on each side of the gap. These logs were herded into another boom and stored in a wide part of the river. The 50 acres of floating logs made a great playground for the local boys out searching for spruce gum. Some of the logs were later pulled out and stored on land for winter use.

The major portion of the drive was released at the appropriate time into a quarter-mile-long channel that would guide the logs down to the dam at McIndoe Falls. The channel was formed by two long booms running length-wise in the river, 25 to 30 feet apart, and held in place by piling driven into the river bottom. The boom logs were connected end-to-end by chains. The channel guided the logs to a 20-foot-wide sluicing gap in the 450-foot-long dam. Releasing the logs over the center of the dam put them into the main current and lessened the problem of stranded logs. To set these upright log piles in the river, the company used a steam-powered pile driver mounted on a large flat boat.

If the river was flowing higher than normal the drive would be held back by the main boom until the flow had receded somewhat. High water would tend to strand many logs in the meadows when the river overflowed the banks. Or the drive boss might hold back part of the drive if the river was too low, a scenario that frequently resulted in logjams. In 1913 the drive was held back because of low water difficulties on the smaller tributary log drives.

The log booms are evident in this early view of the McIndoe Falls sawmill. The channel down the center was used to guide the logs moving on down the river. The sorting gap at the head of the channel led logs into the storage pocket next to the bank for local use.

Once beyond the treacherous Fifteen Mile Falls there were fewer natural obstructions for several miles; dams and bridges again became the principal problem. Shear booms were usually placed at bridges to straighten out the logs in the current and deflect them into the main current.

Commencing at the mouth of the Passumpsic River there was an eight-foot drop in the river over the next five miles of flow down to the next dam, located at McIndoe Falls, where the Connecticut Valley Lumber Co. had a sawmill in the early years. The head of the drive might reach McIndoe Falls in late May or early June, and here the drive was halted behind a strong boom strung across the river. Logs were sorted out for the sawmill, and the main drive was often restrained temporarily if the river flow was too high for proper management of the drive.

McIndoe Falls had some rough spots with strong rapids below the dam where the logs would sometimes jam or "plug" when the water was low. The swift water could jam a few logs so tight that dynamite was occasionally called for.

Log driver Marsh Underwood recalled one jam at this falls when he was delegated to assist with blowing the pile. The "powder monkey" tied some sticks of dynamite onto the end of a log pole, which was then worked under the key logs. The powder man stood behind a large tree, holding the pole, while Underwood was sent out on the pile to light the fuse and run for cover. The explosion tore the pole from the hands of the powder man and threw pieces of log all around them. As Underwood said of the feat, "It went right along with the day's work and there was no honorable mention."

In June 1886 there was a jam of 30 million feet of logs at Milliken's Mill at the foot of the falls at McIndoes, the jammed logs piling as high as 50 feet in some places. Over 200 men worked on clearing the jam. A three-mile-long jam in June 1891 also required 350 men to clear it.

Once past McIndoe Falls, many of the trouble spots that required excess manpower were behind them, thus about half of the 500 or so men working on the drive would be released at about this point on the river. Releasing the men here saved the company a substantial amount of money as board and wages for each riverman was about $3 per day.

Though the major problem spots were now behind the drive, there was still plenty of hard work ahead for the rivermen. In addition to the occasional logjam that might still occur, logs tended to float up onto local farmers' fertile bottomland during high water. Many of these logs would then become stranded against an ob-

Tussles with the Tax Assessor

George Van Dyke and the CVL Co. were frequently at odds with the tax assessor, rightly or wrongly.

1877: Van Dyke refused to pay property taxes in Stewartstown, New Hampshire, because his logging activity was taxed to Van Dyke, Peabody & Co., and Van Dyke attempted a subterfuge by stating there was no such company by that name. As a result of the standoff, the tax collector confiscated one of Van Dyke's horses and sold it for back taxes. The court eventually ruled against Van Dyke in 1881, stating his logging business was a partnership, not a corporation, and that his claim was meaningless.

1878: Asa Smith, general manager of the Connecticut River Lumber co., contracted with a local logger to cut the spruce on some company land in Columbia, New Hampshire, and to pile it on the ice on the Connecticut River for the 1879 log drive. The town attempted to tax the company for the logs, but was denied by the court in 1882.

1901: The company's McIndoe Falls sawmill had been built out over the Connecticut River and the logs were stored within booms in the river. Since the river location was actually in New Hampshire, the Town of Monroe taxed the company for real estate and stock in trade (logs). In 1901, however, the McIndoe Falls mill was almost idle with no logs on hand on the first of April. The efforts of Monroe to tax for $35,000 stock in trade was denied by the judge.

struction or left grounded when the river receded. The damage to the farmland could be considerable.

Landowners were constantly attempting to stop the log drives through court action, but mostly to no avail. Moses Johnson of St. Johnsbury, Vermont, took legal action and in December 1883 received an injunction restraining the Connecticut River Lumber Co. from running logs on the river and maintaining booms at McIndoe Falls. In a hearing on April 1 of that year, before Chancellor Ross, the injunction was dissolved with the court holding that the Connecticut River was a public highway for transporting logs.

Below McIndoe Falls, the four-mile float to Dodge Falls was relatively easy. Dodge Falls, located at East Ryegate, Vermont, is where a paper mill was built in 1905–06, later known as Ryegate Paper Co. A shear boom was used here to guide the logs parallel to the current as they fell into the pitch below.

At the beginning of the log drive some of the supplies, such as

Court Actions Required for Land Damage Settlements

Land damages due to the log drives were frequent but any settlement was infrequent without court action to enforce it. George Van Dyke and the Connecticut Valley Lumber Co. attempted many tactics in an effort to evade responsibilities for damages inflicted. Below are some sample cases.

1887: On or near the first of May, the Connecticut River Lumber Co. placed a boom in the river next to the meadow land of Mrs. Ely Hubbard of Lancaster, New Hampshire. The boom obstructed river flow and backed up 20,000 board feet of logs, many onto the meadow. The farmland remained flooded for two months, spoiling the grass on five acres and breaking off some of the banking. When Mrs. Hubbard couldn't get any compensation from the company she took the matter to court and was awarded $1,000 in damages.

1889: William R. Silver was awarded $1,000 after alleging his land had sustained damage from logs from the annual log drive. But the Connecticut River Lumber Co. contested the award, and it was later determined that the riverbank along Silver's property had broken away previously, and not as a result of the log drive. Damages in the case were thusly reduced to $600.

1888: On May 12, 1888, a large logjam occurred at the McIndoe Falls dam, which forced the dam to give way, resulting in a costly washout of Mill Hill Road in Monroe, New Hampshire, as well as the washout of the Charles McFarland grist mill. The court found that Van Dyke had been negligent in previous needs for dam repair, and the ruling went against the Connecticut River Lumber Co.

1901: When the river floated logs onto the land of Samuel Benton of Maidstone, Vermont, the logs broke down the bank, destroyed trees and were left lying on the grassland. Though local selectmen decided that CVL was responsible for $104 in damages, the company attempted to escape any liability in the case, claiming that the selectmen didn't have the authority to set damages. The court disagreed and set restitution damages.

The bateau was a heavy rowboat with considerable free board, a broad beam narrowing to a high peak and a narrow, pointed bow and stern. The crew normally consisted of four men—two oarsmen, a bowman and a sternman—when doing a chore such as pulling a boom into place. This photograph dates back to 1912. Collection of A. Janion.

Mid-day meals are being delivered by bateau during the 1912 drive. The small man in the bow is Charles Roberts, who was in charge of the floating kitchen. Collection of A. Janion.

food, bedding, tents, tools, grain and baled hay, as well as the cook and his helpers, were carried down the river in bateaux, double-ended 36-foot-long boats that were ideal for maneuvering in the river currents. The boats would also be used to ferry some drivers down to an advance point, take them out to a jam in the center of the river, or to stretch a boom of logs across the river. Mid-day meals were taken by boat to rivermen in outlying river posts. It was a picture of skill to watch the bowman and pilot work the bateau through the river currents.

Indian rivermen from Maine were renowned for their abilities in handling the bateaux, but their record wasn't flawless. In May 1876 four Indians went over Canaan Falls in a boat and capsized. Three were able to make shore, but the body of the fourth was later found under a logjam near the mill dam.

Showing their bravery and skill at "running the dam" are Jack Knights of West Stewartstown, New Hampshire, and John Call of Guildhall, Vermont, in 1909 at the Guildhall dam. One foolhardy riverman proclaimed that he was going to ride every dam on the river while on a log. He was killed at this dam in 1912. Courtesy of Lester Fogg.

The lunch wagon would carry food to the crews working at different points along the river. This photograph was near the mouth of Paul Stream in 1908.

What appears to be the smaller of the kitchen rafts is being propelled by the raft crew with "sculling" oars somewhere above Guildhall. The raft had apparently just been reassembled below a dam because the stove pipe and windows are not yet in place. The crew is taking it ashore to load the stove and supplies.

The upper river "Mary Ann" was apparently hinged in the middle, which allowed it to flop over small dams, as seen here at Guildhall in 1909. However, the raft was reportedly used very little on this upper part of the river. The toll bridge seen here blew down in 1917 and was replaced by an iron bridge in 1919. Courtesy of Lester Fogg.

Mid-day meals were frequently paddled to the crews in the "quick Billy," handled in the stern by the able Henry the Indian in 1908. The young cookee in front is possibly George Drew, a timekeeper from Colebrook, New Hampshire. Collection of A. Janion.

Once the drive reached some roads along the river in northern Vermont or New Hampshire, the supplies were carried by a train of wagons consisting of large four- or six-horse vans that followed along with the drive. The long wagons had CVL painted on the sides, and the team harnesses were trimmed with brass and decked with gaudy tassels.

The company also used a floating kitchen, a large raft with a cooking shed built on top and named the Mary Ann. The term "wanigan" (of various spellings) was applied to the various means of carrying supplies, be it the large raft or the wagons, and was even used for the company store on the woods operations.

There were two kitchen rafts that were used, with the smaller one put in above Colebrook and taken apart at Fitzdale because it

Henry the Indian is pictured loading the canoe with food prepared for the hungry men on the river in 1908. Collection of A. Janion.

A crew of rivermen make use of a log as they sit and enjoy a warm meal just delivered by canoe during the 1908 drive. Collection of A. Janion.

couldn't survive a trip down through Fifteen Mile Falls. A larger Mary Ann was put in the river at Woodsville. This raft would also have to be taken apart at most falls or dams, and the sections were numbered for quick assembly. At the end of the drive the raft sections were shipped back north again by railroad.

WOODSVILLE

The distance from Dodge Falls to the next falls at Woodsville was a short one. Another holding boom was set across the river at Woodsville at a location just above the narrows, up river from the Woodsville–Wells River bridge.

The Woodsville boom was one of the principal holding locations on the river. All of the logs would be held there until the drive up above had passed through Twenty-Seven Islands and arrived at McIndoe Falls. The logs held back by the Woodsville boom would sometimes back up the river for six miles, almost to McIndoe Falls.

The Ammonoosuc River, which carried the International Paper Co. log drive, joined the Connecticut at this point just below the boom. The I.P. Co. drive would usually be held back until the CVL drive went through, but if not, the CVL logs would have to wait. Five to eight men remained stationed at the Woodsville boom. When the logs were let out from the boom they would be released in sections, after men had been stationed down river at the bridges.

The Woodsville boom was also used to hold back the drive until unusually high water had subsided a little, or when word was received about a bad logjam down river. As always, however, high water could create big problems at a boom. In 1906, for instance, the drive broke loose from the Woodsville booms and ended up in a big jam within the narrows near Hanover.

There were also occasions during the drive when a few of the rivermen couldn't resist the temptation to visit a nearby village for the evening, especially if there was a saloon or two in town. It was said that the county jail at Horse Meadow, near Woodsville, was sometimes filled with river drivers after they had caused a ruckus.

The story has been told more than once about a certain Ed Smith who was on a drunken stroll down the Woodsville sidewalk when he spotted a beautiful blonde girl on the other side of the window, with not much more on than her silk hose. With a whoop, Smith leaped through the plate glass front and grabbed her. What he held tightly was a half-nude mannequin standing in the display window of E. A. Sargent's store. Smith had to be subdued by Sheriff Davis, and two others, C. L. Bailey and John Baltis. Smith is said to have drowned the next year on the drive at Perry Falls.

Dry towns were an inconvenience to the lumberjack, but not an insurmountable obstacle. Making the best with whatever was on hand, the thirsty riverman would drink anything in the line of liniment or extract if it had a kick to it. They might clean out the entire stock of patent medicines from the shelves of a local drug store or general store.

Van Dyke strongly disapproved of vandalism or theft by any of the rivermen. He once fired two men on the spot after they had stolen some coats at Hartland, Vermont.

As rough as he was with those who opposed him, Van Dyke was absolutely fair with his men, even compassionate at times. He might fire a man and hire him back ten minutes later. Bill James was once fired for not lacing up his boots—loose laces could result in stumbling while on the logs—but was rehired immediately by Van Dyke.

Fights occasionally occurred among the rivermen, unavoidable with that many men working and living closely together. But they were usually fair fights. Log driver Bill James once had a serious disagreement with a fellow worker, and the next day the opponent

The Woodsville Narrows, where a boom was stretched to hold back the drive. Courtesy W. George Cook.

Many of the rivermen were adept at riding a log.

hit James from behind (and in the back) with a peavey, resulting in the two men both falling off a boom. The foul deed was witnessed by other men who later beat the culprit, and he was never seen again on the river.

In the course of the river drive there usually wasn't much time for recreation, but if there was a lull in the action a riverman might be seen birling or rolling a log out in a quiet stretch of the river. A nimble-footed man who could birl a log while hopefully not falling off was known in some places as a "bubble cuffer." When a man would jump onto a log, whether in play or serious work, he would land heel first so that he could rock onto the log and not catch his toe. With 50 to 60 sharp caulks on the sole of his boots he could get a good grip.

Once the log drivers reached Woodsville they had left many of the hardest-to-work locations behind them and had covered about one-third of the distance to be traveled on the journey down to Mount Tom. The drive had probably grown in volume since its inception up in the river headwaters, joined most years by quantities of company logs cut up in the hills drained by tributaries to the Connecticut River.

9

Log Drives on the Tributary Streams

DURING THE EARLY YEARS, that is, the 1870s and 1880s, the timber harvesting by Van Dyke was primarily around the Connecticut Lakes and the nearby tributary streams, such as Indian Stream and Perry Stream. The cut was heavy, and the cut-over land eventually caught the attention of the newly created New Hampshire Forestry Commission.

In their 1890 report to the state, the commission described the hills on both sides of the Connecticut River as being cut off and eroding badly. The beauty of First and Second Lakes, they claimed, had been destroyed by dams and the resultant backwater or flowage, although the area around Third Lake had not as yet been cut over. The attitude of the woodsmen and local residents was that the forests were inexhaustible. The commission quoted them as stating, "Why, there's hardly anything but woods. We haven't begun on them yet." In 1894 the Forestry Commission reported there were still 150,000 acres of virgin forest in the Connecticut River drainage of New Hampshire's Coos County.

In 1899 Van Dyke moved his crews over to Vermont to concentrate logging on Paul Stream and the Nulhegan River drainage. A spruce budworm infestation had killed or weakened many trees, and the company set up some salvage operations. Adding to the salvage efforts was a 12,000-acre forest fire in Essex County of Vermont in 1903. Blame for the fire resulted in a large lawsuit brought against the Grand Trunk Railroad by CVL Co.

Driving logs down the tributary streams required a technique different from that used on the Connecticut River itself. The normal water flow in these streams, even in the best season, was hardly sufficient to float a long log. And the streambeds were characterized by many turns and an abundance of large rocks. Any hope of floating down logs necessitated a scheme to provide a temporary high pitch or flow.

The rivermen had discovered long ago how to accomplish this feat, however. A series of dams would be built along the streams at appropriate locations where the terrain would allow a sizable backwater to build up. The dams would be built with gates that could be opened at the proper time to release a large flow of water. The winter's sawlog cut would be accumulated at these flowages, either piled on the ice or on an adjacent bank.

Nulhegan Lumber Company Sawmill

Large-scale lumbering activity along the Nulhegan River began with the 1849 construction of a sizable water-powered sawmill by the Baldwin Brothers. The mill was located at the mouth of the river near the village of Bloomfield, Vermont.

The first lumber sawn was formed into a raft and sent down the Connecticut River to Massachusetts in May of 1851. This was said to be the first lumber raft ever sent down along the upper Connecticut River.

In 1880 the faltering mill was revitalized as the Nulhegan Lumber Co. and was rebuilt in 1885 after being destroyed by fire. The steam-powered sawmill was producing 15 to 20 million board feet yearly of spruce dimension lumber.

Large timberland ownership in the Nulhegan River drainage kept 300 men and 150 horses busy during the winter logging season. Logs were driven down the various branches of the Nulhegan River to the sawmill.

George Van Dyke bought the entire operation in 1892, but only operated the sawmill for a short time before shutting it down and moving the machinery elsewhere. What Van Dyke had desired in the purchase wasn't the sawmill, but the thousands of acres of spruce timber along the Nulhegan River that came with it.

The Nulhegan Lumber Co. sawmill at the mouth of the Nulhegan River was purchased by George Van Dyke in 1892.

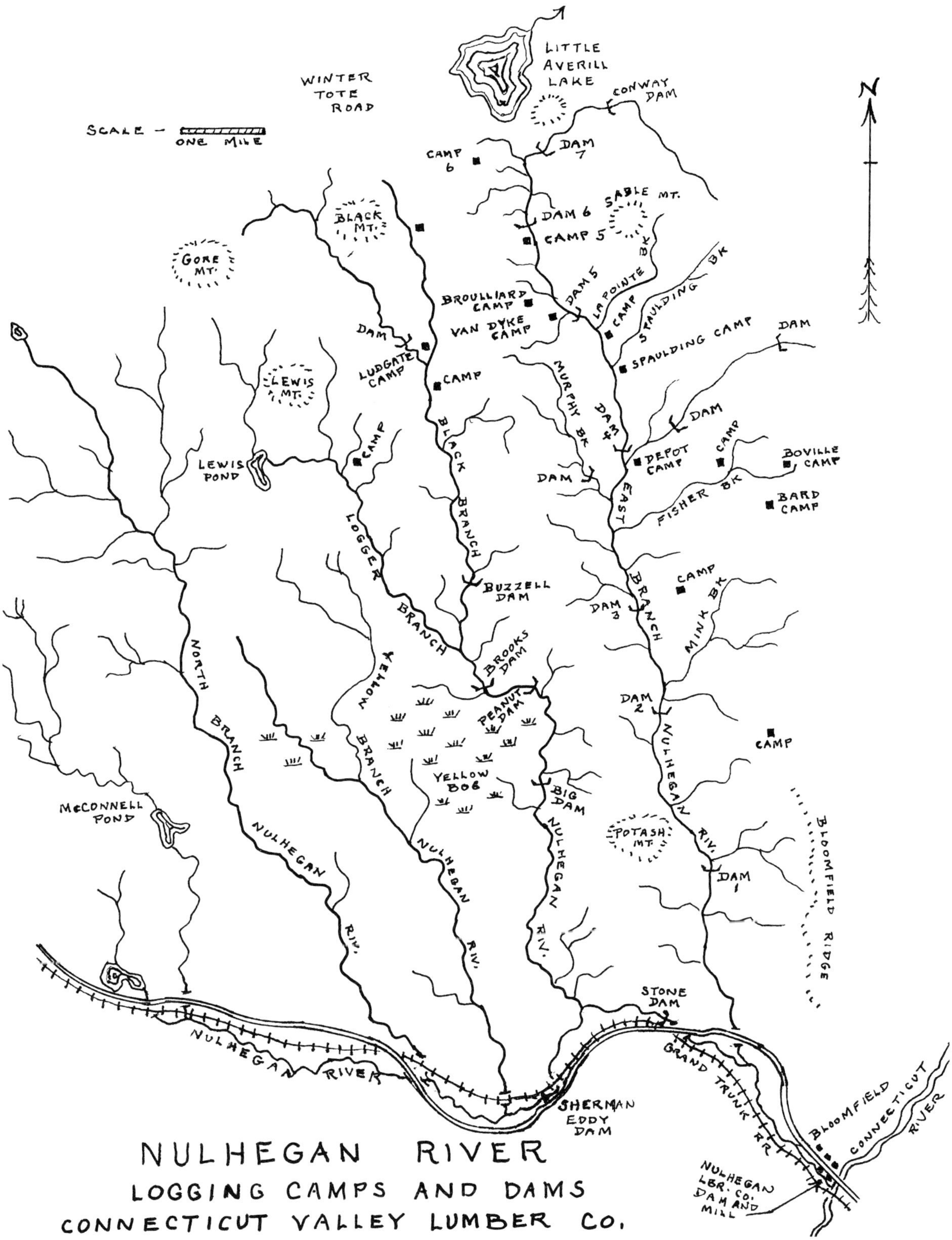

NULHEGAN RIVER
LOGGING CAMPS AND DAMS
CONNECTICUT VALLEY LUMBER CO.

At the appropriate time in the spring the gates would be opened and the stream course flooded with an enormous water flow. The logs in the flowage above the dam were sluiced or guided through the dam opening in order to be carried along in the strong flow. Men with pike poles were stationed at the sluiceway to urge the logs along and to prevent any jamming at the opening. The rivermen knew from experience that when the gates were first opened they needed to hold back the logs for a few minutes until the stream bed had become filled with the churning, foaming water, for a mass of logs being pushed at the head end of the wave did not travel well. As the riverman would say, the logs rode best on the "bubble."

Driving logs down the tributary streams would obviously have to wait until the ice had released its winter grip on the stream, and

The sluice gates were made of heavy planks set loosely on edge in a frame with vertical guides. A heavy pry might be used to raise the gate. The gate would be propped open with a heavy stick that was later cut with an axe when closing the gate. The man on the left is thought to be Phonse Roby.

The gate on this dam had a shaft with a double-ended ratchet arrangement and chains leading from the bottom of the gate to the shaft. One handle held the weight of the gate while the other one got a new hold to raise the gate further.

The small stream bed floods quickly once the gates are opened.

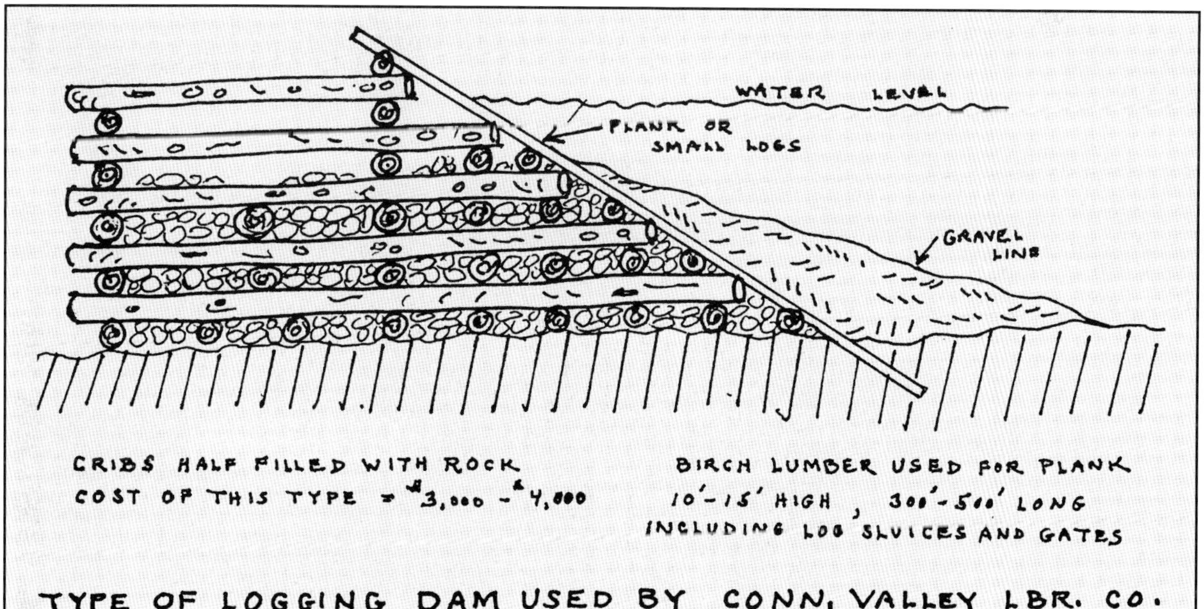

Pictured here is an old roll dam on the river between First and Second Connecticut Lakes. The photo was taken during a period of low water in August of 1914. Stone and Webster report, 1914.

Types of Log Driving Dams

Dams were ordinarily constructed of logs, usually in the form of crib work, and filled with stone.

Sluice Dam: A large dam built of sufficient height to pond a large quantity of water, which could then be released at the appropriate time through gates in the dam that were able to be raised. The logs were accumulated above in the flooded or ponded area and released at the proper time through a gate or sluiceway opening in the dam, guided to the opening by log booms.

Roll Dam: A small dam of sufficient height to back up some water over rock obstructions in the riverbed above. There were no gates; water flowing over the dam was enough to take the logs up over the inclined plane at the top.

Splash *or* Squirt Dam: These were small dams used to hold back a little water for future use. They were often spaced in a series of dams to maintain a water flow for a longer period.

Bank Abutments: Log cribwork or other reinforcements along the stream bank to prevent damage to the bank by the fast moving logs at curves in the stream.

Rugged construction of the sluice gate was needed when the logs came pounding through with the rushing water. The side wall timbers for the 11-foot-wide gate were squared with a wide broad axe, using a burned alder stick for a chalk line. A skilled axeman could make the surface about as smooth as a sawn timber. Note the spillway on the right with flash boards on top to raise the water level of the flowage. Stone and Webster report, 1914.

The up-stream side of the dam was built with an inclined face of planking to prevent damage, as seen in this 1914 view of Perry Dam on Perry Stream. Stone and Webster report, 1914.

An abutment or pier is used to hold the bank during the log drive and to confine logs within the channel.

The rugged construction of the river driver's dam is evident in this 1914 view of the CVL Co.'s Soule Dam on Nash Stream in New Hampshire. The 480 foot dam was built of log cribwork filled with river stone. The 30-foot-long spillway can be noted to right of the three sluice gates; the crest is six feet below the main dam.

Sluicing the logs through the gate at Dam No. 1 on the East Branch of the Nulhegan River, April 23, 1914. Note the long pike poles used by the men on the catwalks and the large ponded area behind. Stone and Webster report, 1914.

the ice break-up usually came a bit later on the smaller streams, especially when the surface was covered with stacks of logs shading the ice. But the rivermen had an answer for this problem; they called it "flushing."

To break the ice, the rivermen would create a series of fluctuating water levels in the stream. The gates at a dam at the upper end of the stream would be fully opened, releasing a full pond of water to rush down the stream course. The next dam below would have the gates closed in order to catch the sudden water flow rushing down. Then this dam would be allowed to release its full pond of water, which would be caught by the next dam below, and so on down to the end.

One gate is raised at Dam No. 1 on the East Branch as the logs have just been sluiced through and the pond drawn down to carry the logs downstream on April 23, 1914.

The ability of a small rocky stream bed to carry 30-foot logs when flooded is well illustrated by this photo of CVL wood in Dead Water Stream, which entered the Connecticut River just above Pittsburg village.

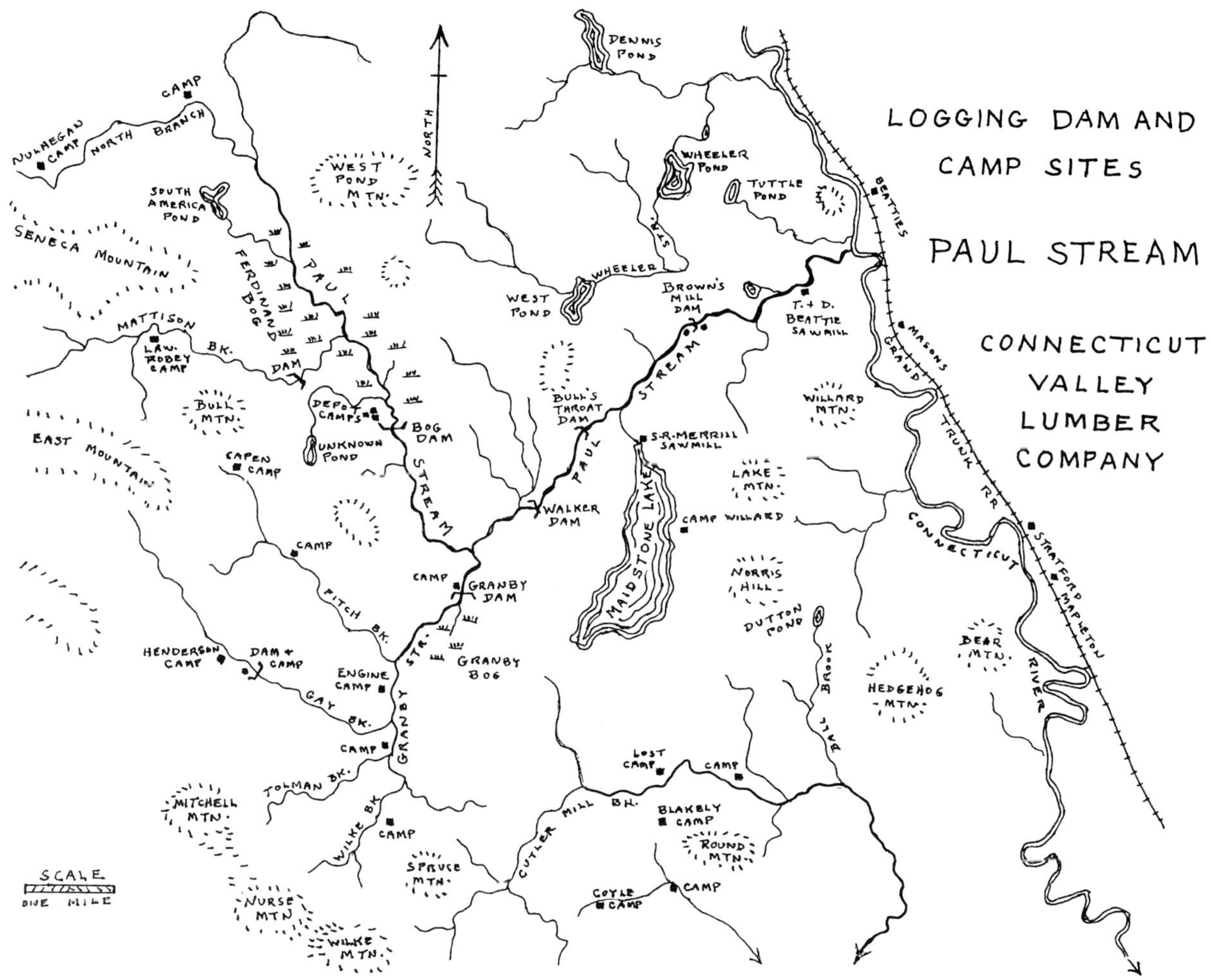

Charles Knight's logging camp on the Nulhegan.

A pile of about a million board feet of spruce rests on the surface of Ferdinand Bog in the upper portion of Paul Stream in 1908. The bog was then flooded to a depth up to about twenty feet by closing the Bog Dam. After the ice was out, a steam donkey on a raft was used to gradually pull down the pile. The stream-driving season only lasted 30 days at most.

To build the large pile the logs were hauled up by means of a cross haul hooked to a pulley block on top of the tall pole in the rear.

The CVL Co. began mechanizing some operations in the early 1900s with the use of a steam-powered Lombard log hauler in the Paul Stream country, pictured here pulling log sleds onto Ferdinand Bog. This is one of the earliest models of Lombard's, which were made in Waterville, Maine. The hauler is being steered by a horse turning the front runners and a teamster sitting on the front of the machine. Later models employed a steering wheel in front over the runners.

The same process might be repeated a couple of times more if necessary, creating the pulsations of extreme high and low water flows. The process seldom failed to break up the ice in the stream and begin the movement of the logs.

This system of flushing the stream would only work during the early spring when there was a high water flow from the melting snows, allowing the ponded area behind the dam to fill up again in a short time. The pond would empty in about 20 to 30 minutes and with a strong stream flow would fill again in two or three hours.

The supply of water impounded by the dams obviously had its limitations of quantity and care was needed to prevent the stream flow from returning to normal before all of the logs were floated down. Coordination was required for timing the release at each dam. Too much water released at once would needlessly exhaust the valuable flowage; too little would create jams on the boulders down below.

The CVL Co. would string a telephone line up beside the stream with a connection at each dam. In that way the drive boss could regulate the water release at each dam at the proper time to keep a good water pitch for as long a time as possible. And should there be a jam, the flow of logs could be stopped. On the Nulhegan River, a flagman was stationed on a high point to signal to the drivers below when a problem was building up.

A large dam location on the lower part of the tributary stream or a sturdy boom across the Connecticut River could hold the tributary drive until the logs could be joined with the main drive on the Connecticut River. Stone Dam on the Nulhegan River backed up a large flowage as a holding location.

Bull's Throat Camp on Paul Stream, August 1914. Stone and Webster report, 1914.

Company-owned timberlands along the East Branch of the Nulhegan furnished many years of log cutting for the Connecticut River drive. Pete Barb was foreman on the East Branch during the era that Bill Smart was drive boss. Large volumes also came out of the Paul Stream basin, and possibly some of the smaller tributary streams. The Ammonoosuc River joins the Connecticut at Woodsville, and here, for many years, the International Paper Co. log drive entered the Connecticut, bound for Bellows Falls. It might precede the CVL drive or come along behind or even on some years be mixed in with the CVL logs.

The flooded area behind Bull's Throat Dam is filled with logs.

The Bull's Throat Dam next to the camp, April 1914. Stone and Webster report, 1914.

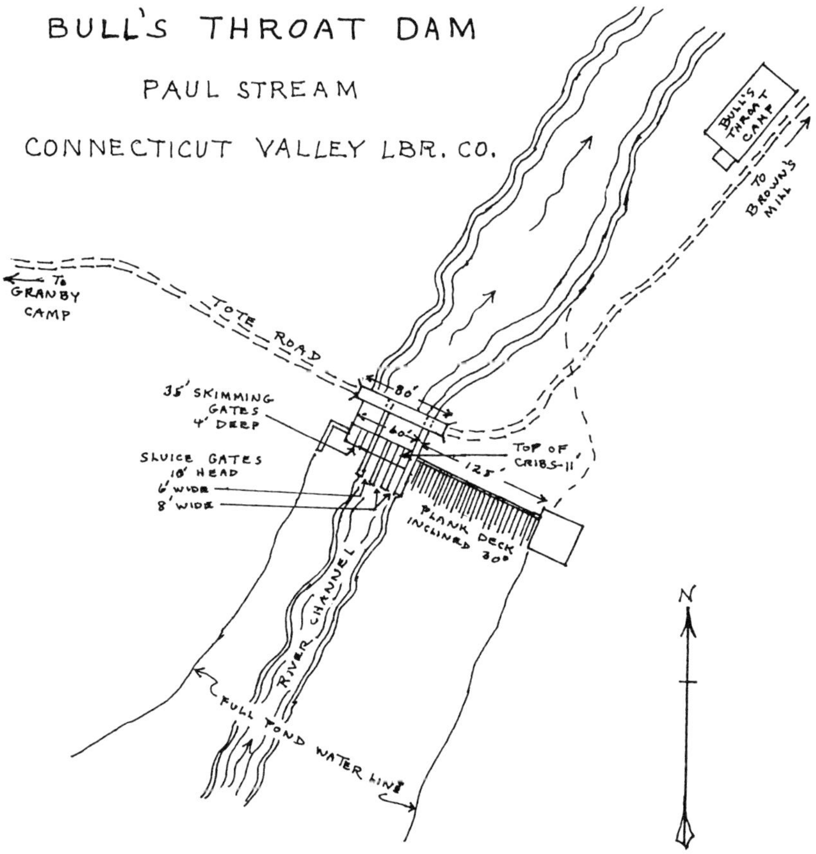

10

The CVL Company Log Drive on the Lower Connecticut River

Woodsville Down to Northampton

AFTER ALL OF THE difficulties encountered along the upper Connecticut River and the occasional loss of comrades drowned in the swirling waters, the river drivers of the CVL Co. had a slightly less toilsome journey ahead on the lower portion of the river. Of the 200 to 250 men working on the crew that had arrived at Woodsville, about 50 were paid off and released at this point.

Gentler currents and deeper waters, characterized by intermittent gradual sweeps and oxbows in the river's course, were now ahead. Men would travel ahead of the drive in bateaux and station themselves at ledges or gravel bars where logs might tend to hang up in the slower current. Two of those bateaux together could carry 25 to 30 men.

Even Olcott Falls, located just above White River Junction, Vermont, at the village now known as Wilder, gave the men no serious trouble. A boom across the river was maintained above the falls to control the movement of the log drive. The boss of the drive might order the boom closed to hold the drive until a good quantity of logs had arrived, or temporarily hold the drive because of a bad jam down river. The boss would also hold up most of the drive until a few released logs had a chance to "wing up" down below along the sand bars, meaning small wing jams which were needed to

form a deeper channel to float the main body of logs. From 600 to 1,200 board feet of logs were required per mile to "wing up" along the river. The boom was easy to close because when open, the logs would pass through an opening only 30 to 40 feet wide at a narrow point in the center of the current; the opening could be closed off with a single log. The boom would usually be kept closed at night, with two or three men on watch.

Logs jam the entire river surface near Hanover in 1912.

Sluicing logs out through the boom opening at Wilder would

The main boom above the dam at Wilder had a narrow opening for logs passing over the dam. The knoll in the background was a frequent camping spot for the wanigan. Note the three men riding a log.

normally require a crew of about twelve men, except during high water, which could pack the logs in tightly behind the boom. A crew of as many as 150 men might be needed in that case. The dead water behind the dam backed up for 12 miles, and thus the logs didn't pile up when pressured.

There were occasional vexing locations yet ahead, however. Just below the mouth of Vermont's Ottauquechee River there were some troublesome rocks in the river, known as the "hens and chickens." The occasional jam on these rocks could be serious. A solid mass of logs might actually create a dam and from this loca-

Olcott Falls (White River Falls)

Just above the mouth of the White River were a pair of falls on the Connecticut River known in the earlier years as White River Falls. Both the lower and upper falls were over 650 feet in length with a descent of an estimated 40 feet. Mills had become established at the falls as early as 1785. In 1810 a canal with locks was constructed on the New Hampshire side to allow the passage of canal boats and rafts around the falls.

An 808-foot cribwork dam was built across the river at the upper falls in 1882, consuming 1.7 million feet of lumber and 3,300 perches of stones. The cost was $50,000. A new concrete dam was built just downstream in 1927, supplanted by another dam built in 1950 three-quarters of a mile down river from the original dam and flooding over the original dam sites.

Paper-making was an industry at Olcott Falls for over 60 years. In 1865 a mill was built on the New Hampshire end of the falls to manufacture paper from straw; the mill washed away in 1872. A ground wood pulp mill and paper mill manufacturing newspaper stock was constructed by the Wilder Brothers in 1883 and sold to International Paper Co. in 1899. The mill operated until 1927.

The New England Power Company now operates a 41,200-kw generating facility at the site.

Olcott Falls in 1882, prior to the building of the dam across river.

tion back up the water as far up the river as the woolen mill at Ottauquechee.

Only a couple of miles further down the river was Sumner Falls in Hartland, a frequent camping location for the rivermen; it also was a very troublesome spot in the river. Ledges protruding out to the center of the river where the river made a turn caused a number of jams and occasional drownings. Some years, it is said, as many as five drownings occurred at this location. Riverman Robert Craig once told of the time he saw a boat capsize when trying to maneuver down below a jam while attempting to set some dynamite.

Areas that once caused extra work for the rivermen at Olcott Falls are now under the quiet waters behind the power dam.

The International Paper Co. mill, pictured here in about 1913, operated until 1927. The dam is visible above the mill and the CVL Co. log boom is pictured above the dam.
Courtesy of W. George Cook.

The main boom at Wilder formed a holding ground for the entire log drive, located just above Wilder Dam. Courtesy of W. George Cook.

Types of Log Booms

A boom was usually constructed of logs fastened together end to end by a chain to form a strong line, but sometimes the boom consisted solely of a strong chain.

MAIN BOOM: Stretched entirely across the river to hold the logs from floating on down the river, often attached to a ledge or pier in the river for extra support. The boom had no effect on the flow of the current. Strong booms were also used to surround and contain a group of logs being towed across a lake. Logs being held in a backwater section of a river would have to be contained by a boom.

High water could cause trouble for a boom across the main current. The logs could "horse up" two or three logs deep against the boom, some of them crosswise, easily parting the boom. A broken boom might enclose a jam and carry it down against an obstruction.

CHANNEL BOOM: Channels were often formed along the river by parallel boom lines. These were used to guide the logs to a dam spillway or to a sorting gap where some of the logs were removed from the log drive. Channel booms were held in position by piers in the river constructed of log cribwork and filled with rock.

SHEAR BOOM: Used to shear or deflect logs away from obstructions or coves by fastening one end of the boom to the shoreline and floating the boom parallel to the river current.

FIN BOOM: Used to guide logs into the main current. The fin booms were unique because they were capable of being manipulated for position in the river. The boom could be positioned with collapsible rudders or fins on the downstream side of the logs. When the fins were extended out at an angle of 30 degrees or more from the log, the current would push against the fins and hold the boom obliquely across the river channel. Some booms were constructed so that a man on the bank could pull the fins back against the logs by means of a rope, allowing the current to push the boom back out of the way for passage of river craft.

A pair of channel booms form a corridor for logs moving down from the sorting boom.

"Harrows" or fins secured to a shear boom would force the boom out at an angle and in a straight line when pushed by the current. Collection of A. Janion.

Shear booms at the railroad bridge at Windsor, Vermont, deflect logs away from gravel bars and into the current.

Napoleon Jacque, seen using an auger, works with other rivermen in "fittin" a boom log during an International Paper Co. log drive along the White River in Vermont. The ends were flattened and a hole bored through with the auger to fit the chain, which joined the ends of the logs.

The boat swamped in the current, and one man drowned. Another man fell off the boom above Sumner Falls that year and drowned before he could be rescued.

Typical of the drownings at this location was that of John Howard in 1873. Howard was helping to free a jam at the rapids a half-mile below the Sumner House when he was thrown from a log. He started swimming for shore but was too exhausted to reach shore and succumbed.

Sumner Falls was a particularly dangerous location for logjams and drownings. Here, the rivermen are attempting to free a dry wing jam during the 1912 drive. Collection of A. Janion.

It's true that the fewest fatalities on the river occurred during the years of water flow that was lower than normal, although such years brought on the hardest work in dragging the logs apart in jams or on ledges and pulling logs off of the shoreline. High water was more dangerous because when a jam broke free the logs tended to drift off faster in the turbulent current and trap the men without an escape route to the shoreline.

A strong chain boom, used to shut off the log flow in the event of the frequent jams below, was stretched across the river above Sumner Falls. A sheer boom was put out along the New Hampshire side of the river to guide the logs into the current and keep them off the rocks.

The Grave of Charles Barber

On the bank of the Connecticut River on the Vermont side just above Sumner Falls, there is a stone slab marking a grave that is inscribed with the name of Charles Barber. The 19-year-old Barber, who hailed from Cherryfield, Maine, had fallen off the logs on June 21, 1895, and succumbed to the fast current at the falls. His fellow rivermen had retrieved his body, carried it into the woods and temporarily covered it with a blanket as word was sent back home to his family by telegram.

Upon learning of his death, his father immediately traveled to Vermont in a wagon and on arrival, took what possessions were in the boy's pockets, plus the $300 that was coming to his son as pay. He then left, but without taking his son's body. Charles Barber's respectful companions scratched his name on a slab of stone, buried him next to the river, and quickly moved on to the next point of danger.

Somebody was kind enough in recent years to arrange for a chain fence around Barber's grave. Photo by author.

Disaster at the Windsor Railroad Bridge

Part of the 1897 logjam at the Windsor, Vermont, railroad bridge has been freed by the rivermen, but the damage had already been done the day before.

On the 8th of June, 1897, a heavy rainstorm spread over the areas drained by the Connecticut and White Rivers, precipitating a large rise in the river levels. The Connecticut River was filled with logs at the time; the boom at Wilder had been opened up a few days previous, releasing 42 million board feet over a span of 20 days. Philo Van Dyke was the drive boss that year.

As the river flow rose to a menacing level, masses of logs flowed southward, gathering in numbers as the wing jams down-river broke free in the rising water and joined the flow. It was customary to allow the wing jams to form on the gravel bars to control the flow of logs. The boom at the Windsor toll bridge was closed in an attempt to hold back some of the large mass of logs grouped together.

The river level rose another three feet on the 9th, straining against any obstruction in the path of the current, and the foreman of the crew stationed at the bridge telephoned up to McIndoe Falls for help. About 30 men under the supervision of Charles Leighton were sent down to Windsor in boats, arriving at 10 A.M. on the 10th. But they were too late; at 5 that morning the boom had broken under the strain of rushing water and backed up logs.

Two thousand feet below the Windsor (Vermont) toll bridge was the large railroad bridge of the Sullivan County Railroad (operated by the Boston & Maine Railroad), and in no time a large quantity of logs was brought to a standstill against the three bridge piers that set in the current. Shear booms installed to guide the logs away from obstruction were quickly wiped out. Within a matter of minutes twelve million board feet of logs were jammed together, covering an area of about six acres. The jam was so thick at the front that the logs went nearly to the bottom of the river and extended high above the surface.

The only unobstructed flow of water was a 60-foot path on the Vermont side. The banking there was soon undermined, however, toppling one of the bridge piers. Along with the pier, a portion of the bridge structure came down, causing damage exceeding $50,000. It took 150 men five days to remove the logjam.

The railroad company brought suit in U.S. District Court in 1898, claiming negligence on the part of the river crew. Van Dyke never paid the claim as directed by the lower court, and the Supreme Court later took up the case in October 1903.

In his defense, Van Dyke claimed that the bridge, which was originally built in 1846–1850 with a wooden superstructure and then renovated in 1889 with the piers raised and the wooden structure replaced with iron, should have been rebuilt with fewer piers, and that one pier was improperly located in the center of the strong river current. Van Dyke also claimed the bridge abutment on the Vermont shore was inferior in construction, and he insisted his crew exercised ordinary care in management of the log drive and brought in a parade of witnesses to testify on this matter.

Van Dyke also argued that the logjam could not have been anticipated by the rivermen. The 1897 flood was indeed a severe one, worse even than the damaging high water of 1896, he claimed. He pointed out that a number of other log booms had been parted along the river, including the ones at Ascutneyville and Turners Falls.

George Van Dyke attempted one of his crafty maneuvers during the course of the trial. At the time of the log drive, the logs actually belonged to a corporation owned by Van Dyke known as the Connecticut River Manufacturing Co. In 1900, Van Dyke attempted to avoid liability and subsequent payment of claims by dissolving that corporation and failing to notify the court, but the strategy didn't succeed.

The court, in its judgment, ruled that the bridge construction was a lawful one and that the Connecticut River Lumber Co. did *not* make reasonable use of the river or exercise ordinary care. The claim was allowed for the Sullivan County Railroad to receive the full amount of damages, which totaled in excess of $50,000.

The wanigan was frequently camped at Sumner Falls for a few days before moving on down river. Tents, when available, provided housing for the cooking crew and some of the rivermen. This wasn't always the case, though. For many, the night's sleep was spent under a blanket in the open or maybe in an abandoned house or old barn. The claim has been made that the rivermen often retired for the night with most of their clothes still on in order to dry them out.

Because of their bunking down in strange places, it is not surprising that the men frequently picked up little critters in their clothes that would make life miserable. In the lingo of the river driver, the man might be seen the next day sitting on the riverbank "reading his shirt." This meant he had taken off his shirt and turned it wrong side out in order to pick off the cooties.

The bridges at Windsor, Vermont, always demanded attention by an advance crew to keep the logs from accumulating against the piers. A toll bridge there was once swept out by a logjam during the early years of the log drives. And then there was the notorious Windsor railroad bridge disaster of 1897, which turned out to be very costly for George Van Dyke.

The arrival of the log drive became a yearly event in almost every town along the Connecticut River. As the word quickly got around that the drive was coming, and as the incessant pounding of the logs might be heard from the nearby village, the townspeople would gather on the banks to watch. If fortunate, they could catch a view of the lead men in the drive, some of them perched on logs passing by in the river current and crouched in a seemingly relaxed pose. Balancing themselves expertly with peavey in hand, the riverman made it look deceptively easy. Spectators would holler and amuse the riverman with such questions as "Do ya ever get your feet wet?"

A big event was in store for the village folks if the rivermen made

Spectators gather on the riverbank at Windsor, Vermont, on what appears to be a Sunday afternoon.

Carrie Pratt's brothel on lower Westfield Rd. in Holyoke, Massachusetts, was well known to the prominent men as well as the working class. Carrie, second from left, would have as many as eight girls residing in the old farmhouse and conducted her operation from 1886 to 1916. Holyokers had a saying for many years, "busier than Carrie Pratt on a Saturday night." Courtesy of Holyoke Public Library.

their encampment near by; this was especially true for the boys. River drivers were in a class all by themselves, always flirting with danger, even death. To the boys and young men these brave men with caulked boots were like gods.

The wanigan might remain camped in one location for a few days, especially if problems were encountered on the river, but it always had to be ready to move down river with the rest of the drive. Among some river drivers, there was a superstition that if the wanigan was not moved far enough down river, thus requiring the men to walk back up for their meals, it became a miscalculation that meant bad luck for the drive.

Whenever there were such large aggregations of men, such as on a log drive, there would surely be a few female followers to provide a service in return for some of the logger's hard-earned money. Historian Stewart Holbrook wrote that only one of the gals left her fame secure in history, an older woman known as "Old

A logjam at the Cheshire Bridge in 1900.

Colorado." She became a fixture for many years, following the drive all the way from First Connecticut Lake to the terminus at Mount Tom. She surely had competition in the larger villages, but in between it was Old Colorado or nothing. The aging gal kept her prices at a reasonable level, never more than a dollar, and frequently only fifty cents.

The Cheshire Bridge at Charlestown, New Hampshire, was another trouble spot that caused a few jams. In July 1892 the drive jammed against the abutments and piled up as high as the woodwork on the bridge. The churning water began to wash away the banks as 110 men worked on the tight mass of logs. Teams were allowed to pass over the bridge at their own risk, but the bridge company stopped collecting tolls. The major concern was that if the jam broke into a sudden rush, the large mass of logs would take along the Fall Mountain Paper Co. pulpwood logs, which were securely corralled in a boom on the side of the Connecticut River just above the paper mill at Bellows Falls.

Other than the occasional highway or railroad bridge, there were no major problem spots below Sumner Falls. None, that is, until the drive reached Bellows Falls.

BELLOWS FALLS

The large, precipitous drop in the Connecticut River at Bellows Falls was the key factor in attracting industries to the area as far back as 1802. Here was built the first commercial canal in the U.S., and the first paper mill to manufacture paper from wood pulp. The Falls also became the site of very conflicting interests for the use of the river water, resulting in disputes which were a contributing factor in the eventual demise of the Connecticut River log drives.

The narrow rapids and falls swing around the village of Bellows Falls.

Logs jam up on the ledges below the Tucker Toll Bridge at Bellows Falls.

The Connecticut River makes a drop of over 50 feet at a sharp bend at this point in the river, tumbling over ledges and down into a large area of backwater. Large crews of men were kept busy at the falls, pushing logs into the narrow river channel after they had tumbled over the dam at head of the falls, and also breaking up the small jams that kept occurring on the ledges.

The head of the drive would frequently arrive at Bellows Falls in the latter part of June, although it might be a month later during the years of a low water pitch. With good water flow and the logs moving rapidly, the CVL Co. drive could pass through the site in as little as 16 days. During the high water in 1910, when there was a flow of three-and-a-half feet of water over the dam, the rivermen passed the logs through in two weeks. But that was unusual.

The "tending out" crew below the Bellows Falls railroad bridge.

Paper Mill at Bellows Falls

First built about 1870, the large paper mill at Bellows Falls was originally known as the Fall Mountain Paper Company, named after the prominent hill on the New Hampshire side of the Connecticut River. The claim has frequently been made that this was the first paper mill in the world to manufacture paper from wood pulp. The Fall Mountain Paper Co. acquired timber holdings in central Vermont and drove long-length pulpwood down the White and Connecticut Rivers for many years.

In 1898, the Fall Mountain Paper Co. was included in the consolidation of 18 northeastern paper mills to form a new corporation, the International Paper Company (IPC). William A. Russell, the principal owner of Falls Mountain Paper Co., was a leading figure in the merger and the first president of IPC.

The pulping process in the mill at that time was both ground wood and sulphite, keeping as many as eleven paper machines busy. The products were cardboards, boxboards and newsprint, and the employee force was over 500.

The International Paper Co. had large annual wood drives down the Connecticut River from company timberlands in the watersheds of Vermont's White River and New Hampshire's Wild Ammonoosuc River. The company drove long-length pulpwood logs down the river until 1915, and then four-foot-long pulpwood until about 1924.

Relations between the mill laborers and management of both the Fall Mountain Paper Co. and IPC mills at Bellows Falls had a long history of bitter disputes and strikes, which occasionally resulted in violence. In August 1886 the workers at the company's steam sawmill struck for wages greater than the $2 per day they were receiving. Their demands were immediately denied, their positions filled with replacements, and on August 26 the former mill employees boarded the 4 P.M. train and left town to seek work elsewhere.

During the reign of International Paper Co. there were continual walkouts, and after 20 years of particularly hostile labor strikes, the company permanently closed the mill in 1926 and sold the property to a new hydroelectric company.

The International Paper Co. papermill at Bellows Falls. The logs pictured are jammed on the ledges below the falls.

The large paper mill at Bellows Falls, originally the Fall Mountain Paper Co., and then International Paper Co. in 1898, also drove 5 to 10 million board feet of pulpwood logs down the Connecticut River each year, beginning in the 1870s. The paper company drive would precede the CVL Co. drive, if at all possible, storing the wood at Bellows Falls within a boom along the Vermont shore until the CVL drive has passed. There were many years when the paper company drive had to be held back by a boom on the White River or the Ammonoosuc River until the CVL drive had all passed through. And there were occasions when high water broke the holding booms on one or the other of these tributary rivers, and the logs mixed together, requiring a tedious log sort at the Bellows Falls boom.

The International Paper Co. drive often arrived at the Bellows Falls destination first, with the logs safely boomed out of the way by the first or second week in June. The CVL drive rarely arrived at Bellows Falls before the last of June, most often arriving in July.

High water sometimes proved troublesome for the mills as they attempted to secure a main boom above the Bellows Falls dam. In 1893, heavy spring rains caused the Connecticut River to run higher than had been known for at least seven years, and the Fall Mountain drive reached Bellows Falls by the second week in May. The river was ten feet deep over the top of the dam, and the paper company had trouble stretching the boom across the river up above. Many of the logs broke free from the boom and were lost over the dam. Retrieval of these lost logs was impractical, and they eventually mixed in with the CVL Co. sawlogs.

The big annual event for the citizens of Bellows Falls was the arrival of the drive and the working of the CVL logs over the dam and down through the falls. The process was a lengthy one and required a good flow of water over the dam to carry the logs, a situation which severely vexed the local industries and their employees. The former boat canal and locks bypassing the falls had by that time been turned into a wide conduit conveying water for industrial use, principally the large paper mill. However, when the CVL log drive arrived the canal water had to be shut off to allow enough water over the dam to carry the logs. Without the canal water, the paper company was forced to shut down their mill for a period of two or three weeks, and sometimes longer, until the drive had passed through. And there was nothing the paper mill management could do about the matter; the CVL Co. had the legal right to use the natural flow of the river, and the logs had to move. One year, it has been claimed, almost 40 million board feet of logs backed up solid above Bellows Falls, reaching four miles up the river to South Charlestown, New Hampshire, waiting to go over the dam.

Legal Case Sets Precedent for Free Passage on the River

An important principle of rights of water passage and use of sluiceways was established in a court case pitting the Connecticut River Lumber Co. versus Olcott Falls Co., dated December 1889.

The Olcott Falls Company and its predecessors had maintained a dam at White River Falls (Wilder, Vt.) for many years, but there was a dispute with the Connecticut River Lumber Co. over passage rights of the company logs.

The legal ruling stated that riparian owners may alter a river channel and divert water for manufacturing purposes, but cannot infringe on the public right to use the river. The defendants were thus required to provide a sluiceway in the dam.

A sluiceway 18 feet wide and 4 inches deep was constructed. It was built with a plank apron covered with iron plates, and with provision for flash boards and for ring bolts to secure booms.

This legal case provided a precedent that was, on occasion, referred to in other litigations and deed covenants.

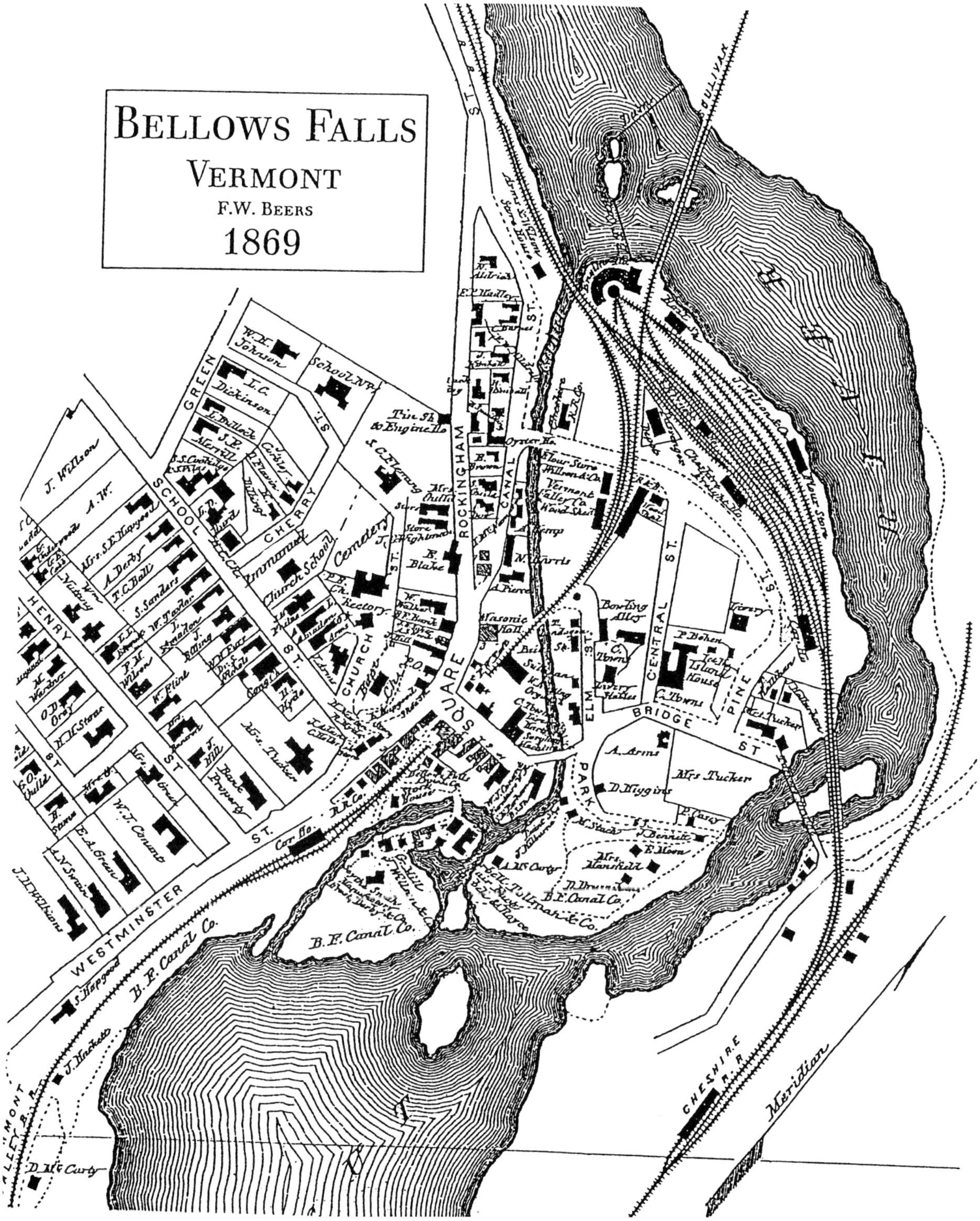
BELLOWS FALLS
VERMONT
F.W. BEERS
1869
ROCKINGHAM
CANAL
SCHOOL
CHERRY
GREEN
HENRY
CHURCH
SQUARE
WESTMINSTER
ST.
ELM ST.
CENTRAL ST.
BRIDGE ST
PARK
PINE
CHESHIRE R.R.
SULLIVAN
B.F. Canal Co.
Cemetery
Meridian

In March 1913, the Sharon boom broke, releasing 2.5 million board feet of logs down the White River. The logs, coupled with the high spring water, wound up dislodging the bridge at White River Junction from its piers.

The International Paper Company Timberlands in Vermont

In 1890, the Fall Mountain Paper Co. of Bellows Falls began purchasing timberlands with virgin growth in Vermont and New Hampshire. After the Fall Mountain Paper Co. was merged into the newly formed International Paper Co. in 1898, land purchases were continued by the IPC subsidiary, Champlain Realty Co. By 1912, the company owned almost 80,000 acres on the watershed of Vermont's White River and over 67,000 acres on the watershed of New Hampshire's Wild Ammonoosuc River.

Company river drives on these two rivers supplied the IPC mills on the Connecticut River at Wilder and Bellows Falls, Vt. These river drives continued until 1926.

International Paper Co. logs tumble over the falls at the Gaysville Dam in 1915, the last year of the long log drives.

It required a brave man to tend out the International Paper Co. logs at the Bath bridge on the Ammonoosuc River in New Hampshire.

An amazing pile of pulpwood logs lie along the south bank of the White River in Sharon, Vermont, in 1912. The logs were apparently dumped there by the railroad above and stacked by the donkey engine seen on the raft at the left edge of the photo.

Long-length pulpwood, all 12 feet long, jam the upper White River in Rochester, Vermont.

The White River log boom at Sharon, Vermont, noted by the piers on the left edge of the photo, held back the I.P. Co. wood until the CVL Co. drive had passed White River Junction. The small sawmill of Sel Noyes, seen on the right, had nothing to do with the huge volume of I.P. Co. wood.

The observer shown here has the stature of the boss as he watches the men working on a boom that is secured next to the dam's gate mechanism.

Some accounts relate the time when the paper mill manager refused to close off the canal until he had completed a rush order for paper. He was approached by either Van Dyke himself or the boss of the drive and was given 12 hours to close off the flow of water through the canal; if not, he was told, the rivermen would be allowed to take matters in their own hands. The canal was soon closed off; the mill's wheels stopped turning.

In 1904 the water was so low on the Connecticut that the water was shut out of the canal from July 11 until at least the end of the month. That denied a payroll of eight to ten thousand dollars per week, and it was reported that the laid-off workers lined the riverbanks mouthing unkind words toward the rivermen. As it so happened, George Van Dyke was in town personally superintending the difficult chore and he did his best to convince the townspeople that his crew was doing the best they could.

Five years later, in 1909, the drive was troubled by low water, head winds and jams on the rocks, and it took 26 days to clear the dam and falls. The strong winds, it was reported, would at times even blow the logs back up the river a short distance. The rivermen hastily built a new sluiceway and blasted out some rocks on the New Hampshire side of the falls, but the troubles continued to tie up 60 men at Bellows Falls that year.

Bellows Falls was the scene of several tragedies over the years, although the first CVL death at the Falls didn't occur until July 1892. Four men went out in a boat that year to drive some iron pins into the dam in order to hold the flashboards. In the boat were Alcomy Rine, Bill Shea, Freeman Judson and Holman Flaherty. They were holding onto a rope attached to a boom above, but the current was so strong they couldn't keep a grip on the rope. Patrick Toole, who was holding the rope at the boom end, got a finger caught in a bight of the rope and, it is said, lost his finger. Flaherty jumped for safety but the other three in the boat went over the dam and were carried under water. Shea and Judson struggled out, but 22-year-old Rine, from Millville, Nova Scotia, drowned. It was his third year on the drive.

In July 1915 a husky riverman walked into the office of Dr. J. P. Lenehan in North Walpole and asked the doctor to do a little something for a long, deep cut in his forehead, put there by the butt end of a log when it dropped on him. The doctor could see that the cut was a serious one and needed stitches; the log driver said no, just some dressing. When told that a bad scar would result without stitches, he replied, "A few scars, more or less, won't trouble me." He then took his pike pole and went back onto the river.

Another problem that surfaced when there was insufficient water flowing over the Bellows Falls dam was the tendency for logs

The Bellows Falls Boom

Log booms were commonly used by all river drivers, utilized to temporarily hold the logs en masse from further progress down the river, to sort out or guide the logs to certain points in the river, or to hold the logs in storage within an area separate from the main flow. Booms were constructed by securing larger logs end to end. Piers, constructed of log cribwork and filled with rock, were set in the river to secure and hold the boom.

On August 23, 1883, a charter was granted to the Bellows Falls Boom Co. (consisting of George Van Dyke et al.) "to erect and maintain booms and piers in and across the Connecticut River from the dam of the Bellows Falls Canal Co. for two miles upstream, also side or branch booms where necessary, and to attach to the bank for the purpose of stopping, sorting or securing logs, masts, spars and other lumber floating down said river and said booms to be provided with sufficient and suitable assorting gaps; not to interfere with the reasonable use of the river as a public highway."

When the local paper company logs were occasionally mixed in with the CVL logs, they had to be sorted out up at the head end of the channel boom which ran lengthwise in the river. The head end was located up near Searles Crossing at South Charlestown, New Hampshire. Sorting the logs was hard work, and if at all possible, was avoided by having separate river drives. But that was not always possible. In June of 1910 there occurred a large logjam at the sorting gap when the logs could not be handled fast enough.

The International Paper Co. pulpwood logs would be boomed and stored in the water within pockets along the Vermont shore about a mile above Bellows Falls, preferably before the passage of the CVL Co. drive. Later the logs were herded across to the New Hampshire side where they were hauled onto shore by a steam donkey engine mounted on a raft. The pulpwood was later cut to short lengths and taken to the mill by truck.

Fast current caused by high water would occasion-

Pulpwood logs are removed from the Connecticut River at Bellows Falls by means of a steam donkey and cables stretched between a mast on the bank and a 65-foot-high tripod on the raft, braced with stiff legs.

to jam in the narrows down below, where there wasn't enough water flow to move them through the rock-strewn channel. Take for example the log drive of 1888. In the second week of July, the advance group of rivermen arrived with the head of the drive and began running logs over the dam. During the weekend they got a half-million feet of logs over the dam before the mills opened the canal gates to start operating on Monday. But then the trouble began. No sooner had the river flow been cut by the opening of the canal than the logs jammed at the narrows. The riverman had to resort to dynamite and before Monday was over had used more than a barrel of dynamite cartridges, damaging many logs. The

ally part a boom. In June 1892 the Bellows Falls boom broke, sending a mass of logs down to Brattleboro, Vermont, where they piled up against a bridge. The Fall Mountain Paper Co. reportedly lost three million board feet of logs down the river, but they were later sold downstream.

The main boom above the Bellows Falls dam is seen here, with the village of North Walpole, New Hampshire, pictured in the background.

The channel boom at Bellows Falls ran down the middle of the river, and was secured by piers. The paper mill logs, seen above on the left, are on the Vermont side of the boom. Courtesy of W. George Cook.

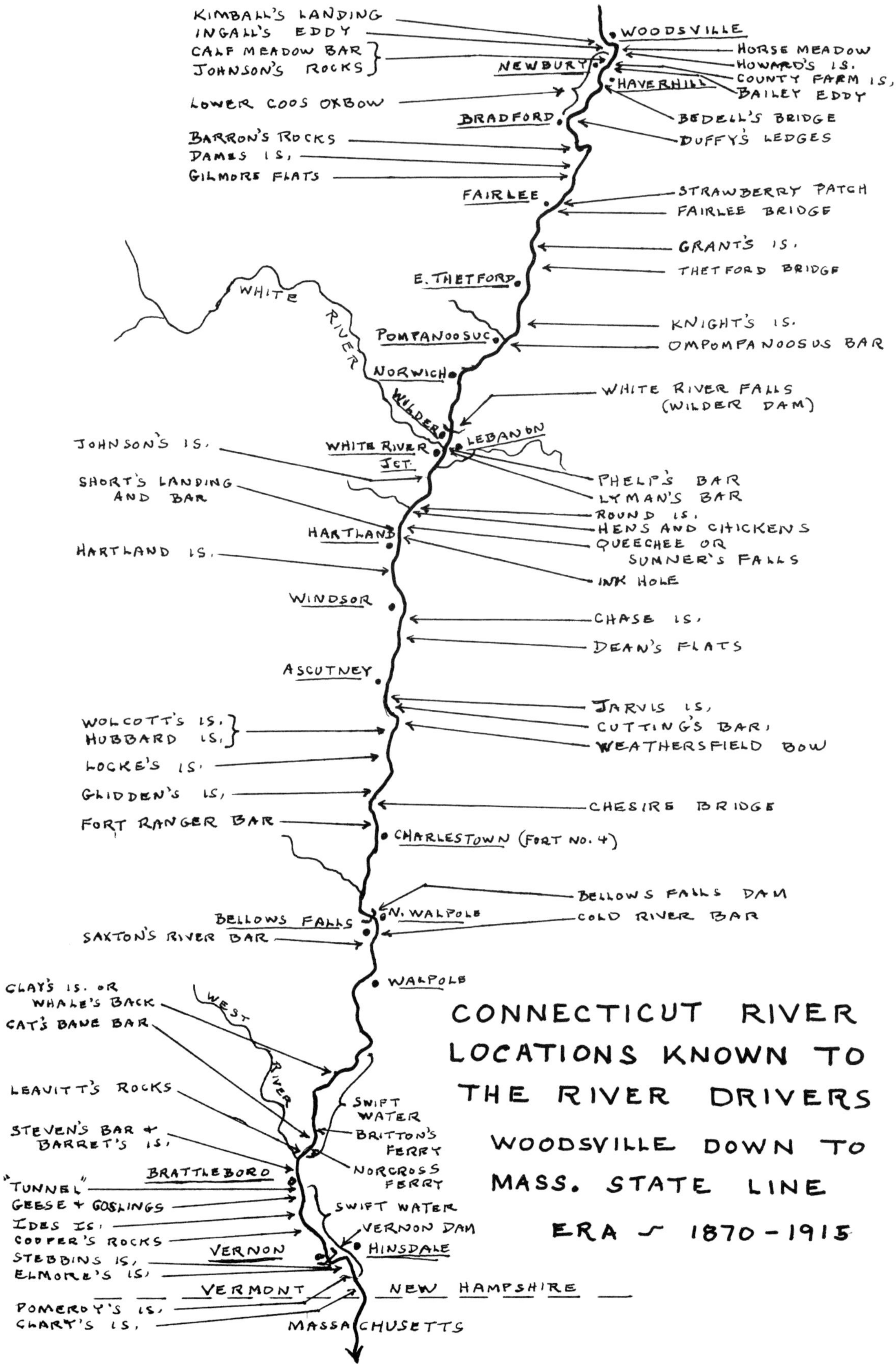
CONNECTICUT RIVER
LOCATIONS KNOWN TO
THE RIVER DRIVERS
WOODSVILLE DOWN TO
MASS. STATE LINE
ERA ~ 1870-1915
KIMBALL'S LANDING
INGALL'S EDDY
CALF MEADOW BAR
JOHNSON'S ROCKS
LOWER COOS OXBOW
WOODSVILLE
NEWBURY
HAVERHILL
HORSE MEADOW
HOWARD'S IS.
COUNTY FARM IS.
BAILEY EDDY
BRADFORD
BEDELL'S BRIDGE
DUFFY'S LEDGES
BARRON'S ROCKS
DAMES IS.
GILMORE FLATS
FAIRLEE
STRAWBERRY PATCH
FAIRLEE BRIDGE
GRANT'S IS.
THETFORD BRIDGE
E. THETFORD
WHITE RIVER
KNIGHT'S IS.
POMPANOOSUC
OMPOMPANOOSUS BAR
NORWICH
WHITE RIVER FALLS (WILDER DAM)
WILDER
LEBANON
JOHNSON'S IS.
WHITE RIVER JCT.
PHELP'S BAR
LYMAN'S BAR
SHORT'S LANDING AND BAR
ROUND IS.
HENS AND CHICKENS
HARTLAND
QUEECHEE OR SUMNER'S FALLS
HARTLAND IS.
INK HOLE
WINDSOR
CHASE IS.
DEAN'S FLATS
ASCUTNEY
JARVIS IS.
CUTTING'S BAR.
WEATHERSFIELD BOW
WOLCOTT'S IS.
HUBBARD IS.
LOCKE'S IS.
GLIDDEN'S IS.
CHESIRE BRIDGE
FORT RANGER BAR
CHARLESTOWN (FORT NO. 4)
BELLOWS FALLS DAM
N. WALPOLE
BELLOWS FALLS
COLD RIVER BAR
SAXTON'S RIVER BAR
WALPOLE
CLAY'S IS. OR WHALE'S BACK
WEST RIVER
CAT'S BANE BAR
SWIFT WATER
LEAVITT'S ROCKS
BRITTON'S FERRY
STEVEN'S BAR + BARRET'S IS.
NORCROSS FERRY
BRATTLEBORO
"TUNNEL"
GEESE + GOSLINGS
SWIFT WATER
IDES IS.
VERNON DAM
COOPER'S ROCKS
VERNON
HINSDALE
STEBBINS IS.
ELMORE'S IS.
VERMONT
NEW HAMPSHIRE
POMEROY'S IS.
CLARY'S IS.
MASSACHUSETTS

The great logjam of 1902 piles up against the bridge at Brattleboro, Vermont, shortly before being dynamited.

following day they began to blast out some of the ledges at the narrows, in the process damaging the toll bridge and nearby buildings as chunks of logs flew about in the air. Lights were hung from the two bridges so that the men could continue to run logs through at night. Eventually the drive moved on.

The company's dynamite supply was once used for a far different purpose. It was July 4, 1914, and the rivermen celebrated by setting off 10 sticks tied onto a plank pushed out into the river. The resulting blast broke windows and plates in homes on the North Walpole side of the river and created a large scare among the residents. Two local men threatened the log drivers with guns and convinced them never again to do such a foolhardy deed.

Starting from the early days on the river, drunkenness became an annual problem in Bellows Falls because of the rivermen's extended stay there. The company tried its best to discourage heavy drinking, paying off and discharging the guilty troublemakers the next day. Stricter rules helped; in 1893 there were only four cases of drunken rivermen.

The log drivers' tent encampment or wanigan obviously had a longer stay at Bellows Falls, and was usually pitched on the New Hampshire side just below the toll bridge. In some years there would be a camp of gypsies set up close by the wanigan. Despite the

Rivermen pick apart the 1902 Brattleboro logjam. The round brick building seen at the far end of the bridge is a gas storage house.

problems incurred by the arrival of the log drive crew, many residents of Bellows Falls benefited by their presence and enjoyed the yearly show. While in town during the 1886 log drive, the rivermen bought $1,200 in groceries, $400 in meat and $1,000 in clothing from the local merchants.

After the log drive passed through the difficult Bellows Falls, the course ahead was again somewhat easier and the manpower need became reduced in numbers. In 1890, for example, 80 men were paid off there, leaving only 40 to continue on down the river. The men earned $2 to $2.50 per day, payable at the end of the drive or when dismissed; the payroll at Bellows Falls might be as much as $12,000. Quite understandable was the practice of many, upon being discharged, to immediately purchase a new suit of clothes, hat and underwear.

As the log drive floated down into the more populated areas

The rivermen struggle to clean up the stranded logs below the dam at Turners Falls. Note the man on the capstan pulling a small boom of logs across the backwater. The bateau seen here is being transported around the dam on a wagon frame. The Turner's Falls Lumber Co. sawmill is pictured on the other end of the dam.
Courtesy of Peter Miller.

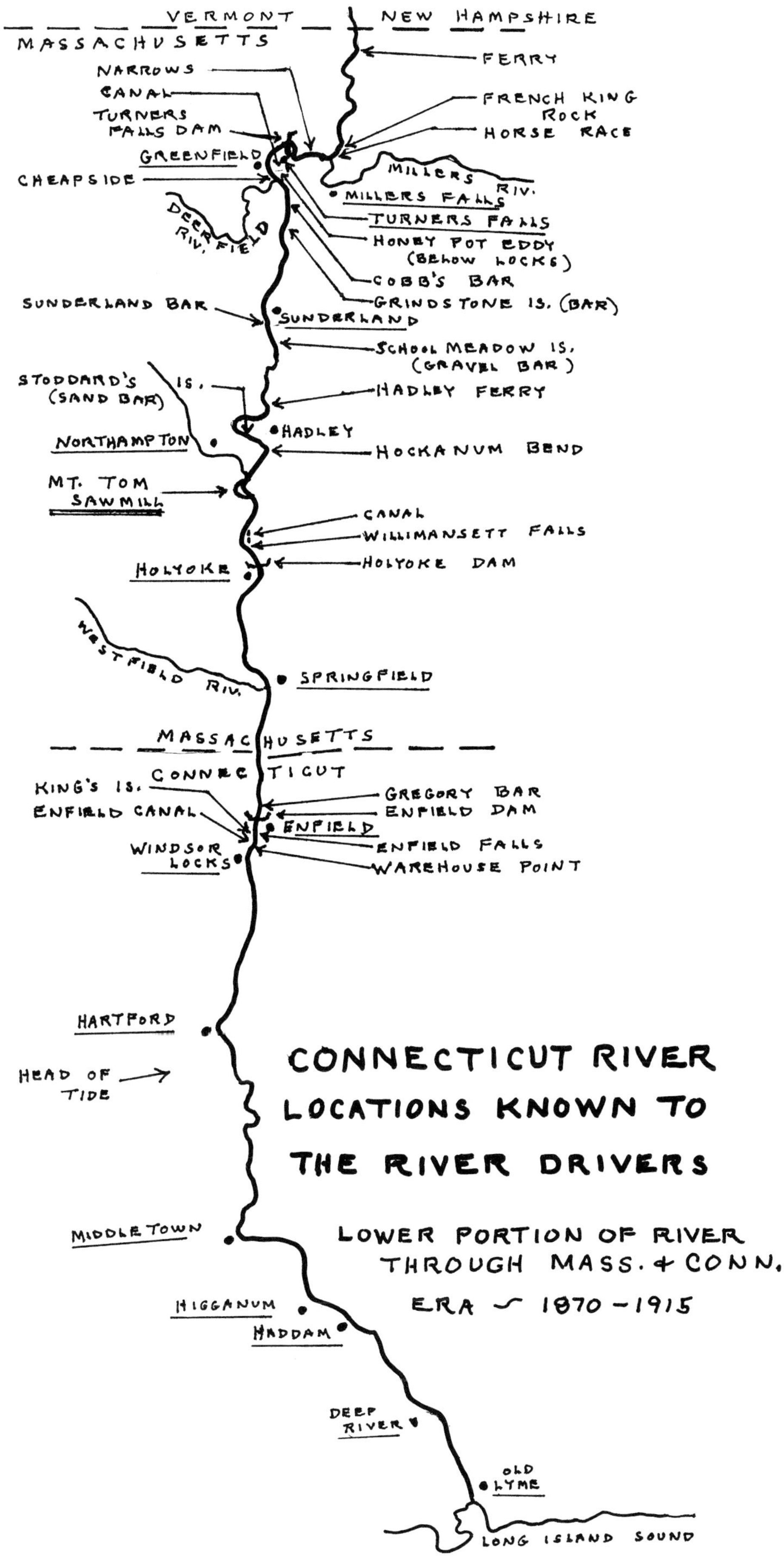
VERMONT
NEW HAMPSHIRE
MASSACHUSETTS
FERRY
NARROWS
CANAL
FRENCH KING ROCK
TURNERS FALLS DAM
HORSE RACE
GREENFIELD
MILLERS RIV.
CHEAPSIDE
MILLERS FALLS
DEERFIELD RIV.
TURNERS FALLS
HONEY POT EDDY (BELOW LOCKS)
COBB'S BAR
GRINDSTONE IS. (BAR)
SUNDERLAND BAR
SUNDERLAND
SCHOOL MEADOW IS. (GRAVEL BAR)
STODDARD'S IS. (SAND BAR)
HADLEY FERRY
HADLEY
NORTHAMPTON
HOCKANUM BEND
MT. TOM SAWMILL
CANAL
WILLIMANSETT FALLS
HOLYOKE
HOLYOKE DAM
WESTFIELD RIV.
SPRINGFIELD
MASSACHUSETTS
CONNECTICUT
KING'S IS.
GREGORY BAR
ENFIELD CANAL
ENFIELD DAM
ENFIELD
WINDSOR LOCKS
ENFIELD FALLS
WAREHOUSE POINT
HARTFORD
HEAD OF TIDE
CONNECTICUT RIVER
LOCATIONS KNOWN TO
THE RIVER DRIVERS
MIDDLETOWN
LOWER PORTION OF RIVER
THROUGH MASS. + CONN.
ERA ~ 1870 – 1915
HIGGANUM
HADDAM
DEEP RIVER
OLD LYME
LONG ISLAND SOUND

along the Connecticut River there would be a number of small ferry crossings to contend with, and these operators also had to give way to Van Dyke's log drive. As soon as the first logs or "best travelers" appeared, Ashley's Ferry, as with many other ferries, had to put a man aboard to push away the threatening logs. They also had to stop their night runs. Within a short while, the ferry had to stop completely when the river became filled with logs.

The mild rapids down through Vernon, Vermont, presented no problems until a hydroelectric dam was built there in 1908. After that, the backwater created by the new dam slowed down the log flow and increased the workload of the rivermen.

The Turners Falls Lumber Co. of Turners Falls, Massachusetts, also put logs into the Connecticut River for the annual log drive. The logs were harvested on land the company owned at the top end of the Connecticut. Their logs, however, were usually mixed in with the CVL logs, requiring the logs to be sorted out at the head end of a center boom above the Turners Falls dam. The Turners Falls logs were stamped with a diamond-shaped mark, the CVL logs with an arrow. In 1893, for example, there were 15 million board feet of Turners Falls logs that had to be sorted out of the drive, and even more in other years.

The Turners Falls Lumber Co. sawmill was built in 1868 at the north end of the 1,000-foot-long, 35-foot high timber cribwork dam across the Connecticut. The mill burned down in 1904. The company also operated sawmills in the North Country of New Hampshire, including a new mill they built in South Lancaster, New Hampshire.

The former flat boat canal and locks at Turners Falls was also used for supplying water for turning the turbines of some local industries, including a paper mill. But when the log drive came through and the water was low, this canal also had to be closed off.

As the main body of the log drive entered into the Commonwealth of Massachusetts, the long trip was nearing completion; the final destination was close at hand. But during the headlong descent of the long river, many logs were left behind, stranded on the gravel bars, ledges and meadows. Thus along behind the main drive came another small army of seasoned rivermen, salvaging the valuable logs forsaken by the river currents. They were said to be "picking up the rear."

11

Picking Up the Rear

ONE OF THE MOST toilsome duties on the log drive was the labor at the rear end of the drive. Logs would become stranded and left high and dry at many locations over the long length of the log drive; these logs were valuable and had to be returned to the river. The entire log drive, including the stragglers, would extend over a distance of 75 to 100 miles along the river.

It was inevitable that some logs would be left behind. During low water the ledges and gravel bars would either be exposed or lie concealed under shallow water and would capture many lone travelers, especially as summer arrived and the river level dropped even more. On the other hand, during high water the situation was different, maybe worse. The river would overflow the banks, inundating the meadowlands and creating ponded areas that the wayward logs would manage to seek out. When the water began to recede from the meadow areas, the logs soon became grounded with no watery escape route. There were some years when as many as four million feet of logs had to be retrieved by the large crew bringing up the rear.

Moving these valuable logs back into the river again was the task of the large crew on the rear, which included as many as 160 men, a head cook and six or seven helpers. During the 1897 log drive, Win Schoppe bossed the rear down as far as McIndoe Falls, and John Stapleton took charge of the remainder; Philo Van Dyke happened to be the drive boss that year.

Moving the logs over ledges or dry land required horse power, and a large contingent of horses would have to travel down the Connecticut River with the rear end crew. The horses traveled on log rafts, sometimes 6 or 8 abreast in two rows on a raft, five feet

With the waters of the Connecticut River having subsided considerably by late in the spring 1910, many logs are left stranded below the Guildhall, Vermont, bridge. Courtesy of Lester Fogg.

Choice riverside meadows, like the one pictured here, might be covered with stranded logs.

An eddy with slow-moving backwater could easily trap logs. A large eddy like the one seen here at Windsor, Vermont during the 1912 log drive, might hold back 8 to 10 thousand board feet of logs, but would sometimes clean out overnight with a change in the water current. Collection of A. Janion.

Horse rafts were made of logs sturdily lashed together.

The horses seem perfectly calm on their 1912 journey down the Connecticut River. Collection of A. Janion.

Both 8-horse and 16-horse rafts were used during the 1914 drive.

apart, and rarely showed any fear. But the trip was not always a joy ride for the horses. In July 1906 four horses were somehow knocked from a raft and drowned. Suspended ferry cables across the river, such as the one at Rockingham, Vermont, were another hazard. Bill James once recalled the time when all of the horses on his raft were swept into the river by a cable and were almost helpless because they had been tied together head to tail. Bill jumped on the back of one horse and cut it loose, but some of the other horses unfortunately drowned.

Once, when the rear crew was working at Dog Island, a little above Bellows Falls, they worked late to finish on a gravel bar and it became dark. They had a 16-horse raft with 12 oars, and upon

Many of the needed supplies, such as bedding and tents, were carried downriver by bateaux like the ones pictured.

A blacksmith's anvil and other equipment can be seen in the bateaux shown here.

their return hit a ferry wire. Eight horses were swept into the river by the cable and four drowned because they were unable to swim properly while still harnessed together.

The large army of men on the rear was divided into two forces. What was called the "rolling rear" was a crew to primarily clean up the logs along the river edge. Several miles behind would come the "floating rear" with the horses to retrieve the difficult logs. There could be as many as 75 horses traveling down the river. A floating "wanigan" with a cook and seven helpers traveled with the floating rear.

Logs hung up on high ledges had to be pulled down to the water's edge by means of a horse and a snatch block, and those logs stranded on the gravel bars had to be dragged off by horse power into water deep enough for the log to float. Some of the gravel bars

Some of the large gravel bars along the Connecticut River could accumulate many logs, requiring many hours of work to free them. Collection of A. Janion.

The teamster and the rigging swinger have brought the team into position for the maul man to set the dog into the log. After the log is in water deep enough to float, the dog would release when hit on the back end. Collection of A. Janion.

The dog is set in the log end by a hefty swing of the maul in this scene from 1908. Collection of A. Janion.

After a few days in the water, horses used during the drive would get cracked skin from prolonged immersion, and their hair would come off. Collection of A. Janion.

were surrounded by deep water, but many of the horses would readily swim out to the bar. While pulling a log off the bar they might have to continue out into deep water, but they were trained to turn and swim back at the teamster's command after the log had been freed from the skid chain with a quick pry of the peavey.

Dragging back the logs stranded on the still-wet meadowlands would obviously cause damage to the choice river-bottom farm-

This old-timer was from Bangor and in respect for his many years on the river, he was given a less strenuous job. He was the "tripper," whose job was to trip free the dog on the next hitch of logs. Collection of A. Janion.

Logs stranded high on a ledge required a long rope, maybe even block and tackle, to apply horse power. Note how battered the logs became during the long river journey of 1912. Collection of A. Janion.

lands. Oftentimes the damage was considerable. This was a constant source of contentious grief for Van Dyke and his field bosses. A number of lawsuits were filed against the company and against Van Dyke personally, and most of the landowners were openly hostile to the rivermen.

Historian Robert Pike passed along the account of the famous Craggie incident in Groveton, New Hampshire. In this case, a sudden freshet and rise in the river level had left a large number of logs on Craggie's meadow, and as such they could not be legally

Townspeople, seen in the upper left as they gather alongside the river, are dressed in their Sunday finest as they watch the rivermen at work during the 1912 drive. Collection of A. Janion.

The wanigan "Mary Ann" dispensed four meals daily to the hungry rivermen, along with salt pork for their blistered feet. Collection of A. Janion.

touched without the owners' consent. That fine point of the law didn't deter Van Dyke; he quickly moved in about fifty men with peaveys and horses and began to drag the logs into the river. At this point the farmer appeared with the sheriff, and work came to a standstill as the men were waiting to see what would happen. Van Dyke reportedly berated his men for stopping, told them to keep rolling the logs into the river, and added that if Craggie or the sheriff said anything, they should roll them into the river, too. The logs went in.

Retrieving the logs from the still waters of an eddy required a different tactic; an eddy boom might have to be employed to surround the logs and sweep them out into the current. A special boom raft was made of squared timbers fastened end-to-end and which when released, extended out to form a convenient boom for this purpose. Afterward the string of timbers was folded up again into raft form.

A raft of supplies for horses and men makes its way down the Connecticut River in the summer of 1912. Collection of A. Janion.

The Mount Tom sawmill is seen here in the center of the photo, where the tall smokestacks are visible. The mill was located on the great oxbow of the Connecticut. It appears that the oxbow is almost full of boomed logs and huge piles of logs can also be noted on two sides of the cove. Courtesy of Holyoke Public Library.

CONNECTICUT VALLEY LUMBER CO.,

MANUFACTURERS OF

LATH, SHINGLES,

SPRUCE LUMBER

—AND—

SPRUCE PILES.

MILLS AT

HOLYOKE, MASS. MT. TOM, MASS.
McINDOE FALLS, VT.

HOLYOKE OFFICE, 149 HAMPDEN STREET.

Mount Tom Sawmill

In 1873 the McIndoe Falls Lumber Co. of Vermont, composed of Stephen Barker, D. N. Eastabrook, Isaac Bowman and R. G. Hinman, built a large steam-powered sawmill on Hockanum Island in the Connecticut River at Mount Tom, Massachusetts, just south of Northampton. The 130-foot by 50-foot mill was capable of sawing 10 million board feet annually.

During the spring of 1873 the McIndoe Lumber Co. drove down eight million feet of spruce from the North Country to get the mill started. At Mount Tom, the logs were stored in the oxbow of the former main riverbed.

A sulphite pulp mill was later built next to the sawmill to process pulp from the slab and edgings residue generated at the sawmill; the slabs were moved over to the pulp mill by a chain conveyor.

In 1878 the company was reorganized into the Connecticut River Lumber Co. with George Van Dyke one of the partners; a later corporate change for the mill renamed it the Connecticut River Manufacturing Co. After George Van Dyke acquired majority control in 1898 the business was reorganized as the Connecticut Valley Lumber Co.

A $40,000 renovation in 1906 installed two new band saw head rigs, increasing production to 100,000 board feet daily, filling about twelve freight cars each day. Manager of the mill at the time was James W. Reid. Harvey A. White later succeeded him.

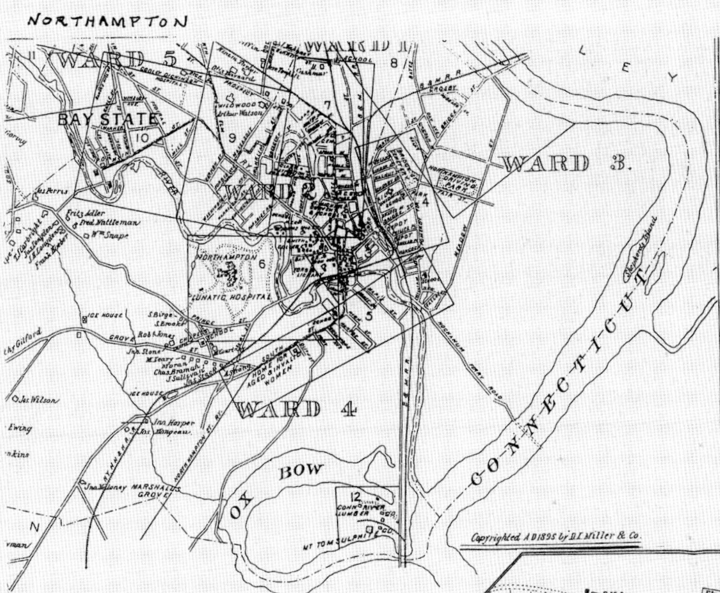

An 1895 tax map shows the location of the company's Mount Tom sawmill at Northampton.

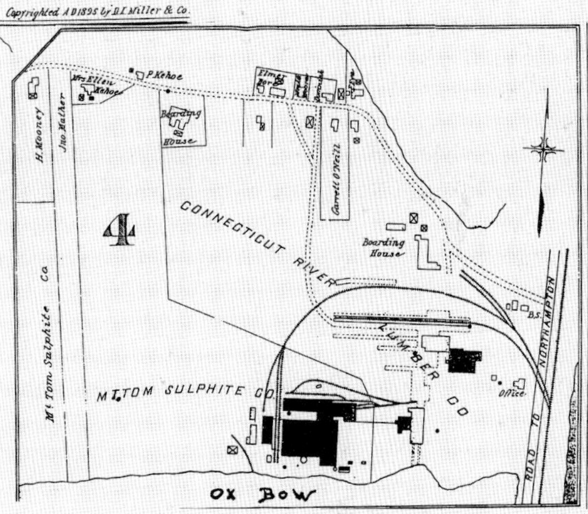

The following year, on April 22, 1907, the Mount Tom sawmill suffered a fate so common with sawmills—total destruction by fire resulting in a $100,000 loss. The Mount Tom Sulphite Pulp Co. mill remained unharmed, however. The 125-man labor force was out of work, and the five million feet of logs on hand were floated down the Connecticut River to be sawn at a mill in nearby Holyoke.

The Mount Tom mill was rebuilt, and in 1908 the sawmill at McIndoe Falls, Vermont, was closed and the machinery brought to the new mill. The main mill building was 220 feet by 56 feet. There were two band saw head rigs, each sawing 48 to 60 thousand board feet per day of spruce. There were over 150 employed at the sawmill, including the yard men and rivermen, and 300 employed at the pulp mill. Many were Polish immigrants.

In 1916 the mill began experiencing protracted labor troubles, which led that November to a cease in lumber manufacturing operations, and eventually the complete closing of the plant the following April, ending 44 years of continual operation.

A jack ladder, used to convey logs into the CVL sawmill at Holyoke, can be seen at the upper right. On the left, logs are being hauled out by a steam donkey for storage on land. The Mount Tom Sulphite Pulp Co. mill is in the background. Courtesy of W. George Cook.

The cook had absolute domain over his crew. The older man in the middle with a mustache is Charles Roberts, captain of the "Mary Ann." Collection of A. Janion.

For the greater part of the existence of the Connecticut Valley Lumber Co., the final destinations of the logs coming down the river were the sawmill sites at Mount Tom, located at the oxbow in the river just below Northampton and at Holyoke, Massachusetts. Most years there were a few million board feet of logs intentionally left at certain mill sites en route, such as Van Dyke's sawmill at McIndoe Falls, the International Paper Co. mill at Bellows Falls, or the Turners Falls Lumber Co. sawmill. But the majority of the long spruce logs came all the way down to be corralled at Mount Tom or Holyoke.

The distance through Massachusetts wasn't long, and there were very few trouble spots. Even the bridges over the Connecticut at Sunderland and Northampton were usually no obstacle to the slow-moving logs. But the renowned river boss Al Patrick once recalled that one of the most exciting experiences during his many years on the river was the time they averted a catastrophe at the Sunderland railroad bridge. Patrick and many of his rivermen were already at Mount Tom in the summer of 1904, awaiting the drive which was about to arrive, when a sudden freshet and consequent rise in the river flow caused the boom at Turners Falls to break, sending twenty million feet of timber down in a solid mass.

Patrick and crew rushed up from Mount Tom and were able to position themselves at the center pier of the Sunderland bridge, where they toiled from 6 P.M. to midnight to prevent stoppage as the logs bore down on them.

Secured to Shepherd's Island a short distance above Mount Tom, there were three booms that ran lengthwise down the river, forming two channels for the logs to pass on into the holding booms at Mount Tom.

The end of the long drive seldom had problems, but such wasn't

Holyoke Sawmill

Sawmills that had been built in Holyoke, Massachusetts, on the Connecticut River at an early date were consolidated into ownership by the Connecticut River Lumber Co. in 1879. Within a short time, just one of these sawmills remained active, that being the large mill a short distance above the Holyoke Dam. This sawmill eventually came into ownership by the Connecticut Valley Lumber Co. and was closed down about 1907.

Local youngsters pay a visit to the Holyoke mill to collect waste wood for home fuel use. Courtesy of Holyoke Public Library.

The Connecticut Valley Lumber Co. sawmill at Holyoke, Massachusetts, can be seen to the left of the tall smokestack. Courtesy of Holyoke Public Library.

Crew members at the Holyoke mill take a break from loading spruce lumber onto railroad flatcars. A wagon load of bundled lath is about to depart. Courtesy of Holyoke Public Library.

the case in June 1897, the year of the Windsor railroad bridge disaster. A sudden large rise in the river flow suddenly pushed the main body of logs down onto Mount Tom before the boom was in place and the logs rushed on by. A boom at the Holyoke dam held for a while, but the wild rush of logs eventually proved to be too much. It appeared that the whole drive might end up in Long Island Sound, but Van Dyke, who had rushed up from Boston, was personally out in a bateau in the river, bellowing and gesturing directions as the men were finally able to close the boom.

Another boom was stretched across the river at Hartford, Connecticut (preceded by the drowning of two rivermen at Enfield as the crew was hurrying down the river to Hartford). Most of the logs were stopped at Hartford, where the company set up a large portable sawmill to process the logs. However, the Hartford boom also broke at one point and some of the logs did eventually end up in Long Island Sound. Van Dyke offered a dime per log for all the logs salvaged and landed at a shore location, where they were collected as the local fishing schooners towed them in. The logs were all boomed together by Van Dyke's men, unaccustomed to working in salt water, and then towed back up the river to the temporary sawmill at Hartford.

With the annual log drive completed at Mount Tom, the logs were all secured in a tight boom within the oxbow, many to be hauled out and piled on shore for winter use. The remainder of the river drivers were paid off and they drifted off to their home farm, some to await the opening of the logging camps for a winter season of lumbering in the deep north woods.

After the crew picking up the rear had arrived, their stock horses were either sold, shipped back by rail, or walked all the way back north again. It was a 12-day journey to walk from Mount Tom back to Littleton, New Hampshire, with the horses hitched to a two-wheel rig to carry their oats and hay. Only trusted teamsters were given this job of returning the horses.

The delivered cost of the logs in 1900, according to the CVL, was $10 to $12 per thousand board feet to cut and pile riverside and another $2 for the river driving expense.

George Van Dyke maintained that the Connecticut River was more difficult to drive than any other river in the Northeast. It was the only river with no reservoirs or big lakes, and thus the rivermen had to take the water as it flowed from the sources with no storage breaks between storms. There would be a freshet above, and the log drivers would not even be aware of it until the river suddenly rose.

12

Van Dyke's Tragic End and the Sale of the Company

GEORGE VAN DYKE was a formidable figure, whether among the timber or along the river. He might be found at any location, his booming voice overreaching the strained voice of the man on the peavey or the harsh commands of the boss of the drive. He knew his men and knew what could be expected.

If all of the tales and legends about George Van Dyke were true he would have had to have been almost everywhere at the same time. He was certainly a man who got around and stood with the men. But many a tale about Van Dyke is no doubt exaggerated, with a few beyond the realm of belief.

As is typical of the timber barons of that era, Van Dyke knew how to drive a hard bargain and how to profit at the expense of those not as astute or perceptive. He built a large empire, ranking high on the list of the New England timber barons. By the turn of the century his Connecticut Valley Lumber Co. owned about 330,000 acres of timberland in northern Vermont and New Hampshire, and his sawmill at Mount Tom, Massachusetts, was sawing from 40 to 80 million board feet yearly. Among his other accomplishments by this date was ownership of the Moose River Lumber Co. in Lowelltown, Maine, and the presidency of Brompton Paper Co., Brompton, Quebec. Ever since his ascendancy to the position of president of the Connecticut River Lumber Co., in 1886, Van Dyke maintained an office in Boston, at 89 State Street. He resided at Young's Hotel until 1908, when he acquired a home in Lenox, Massachusetts.

Reaching the pinnacle of success, however, took its toll on Van Dyke's health. In May 1908 he was hospitalized in Boston, unconscious for three months with "water on the brain," it was said. He recovered but was never again the same robust titan.

In 1909 Van Dyke was back on the river again with the annual spring log drive. This time, however, he was not hurrying up and down the road along the river behind his favorite team of horses; he was not pacing the riverbank. Van Dyke was confined to traveling along the Connecticut River in his red 1908 Stevens-Duryea touring car, driven by his chauffeur, Frederick "Shorty" Hodgdon. He would not relinquish the personal attention he gave toward his wealth of floating timber, but now he had to view much of the activity from the seat of his automobile.

That 1909 log drive, containing 36.2 million board feet of CVL logs, not only went down in the record books as one of the more difficult drives, but also as the one which claimed the most famous fatality in company annals.

By June 3, 1909, the International Paper Co. drive had already reached Bellows Falls and IP's logs were piled on the bank alongside the river; the CVL Co. drive of 36 million board feet was still being held behind the boom at Wells River. Released from Wells River on June 24, the drive took almost three weeks to reach Bellows Falls, and there the men had trouble getting the logs over the dam due to contrary headwinds and logjams. In all, it took the rivermen 26 days to get the drive through Bellows Falls.

Another week or so brought the log drive to Turners Fall, with George Van Dyke watching the labors of his men as best he could in his auto confinement. On Sunday, August 8, at the order of Van Dyke, his chauffeur parked the car close to the edge of a 75-foot high riverbank at Riverside, on the east side of the Connecticut. Van Dyke watched the men below as they worked the logs through the rapids at Turners Falls.

Satisfied with the rivermen's progress, Van Dyke then told his driver, "Shorty" Hodgdon, to drive him back to the Welden Hotel in nearby Greenfield. What happened next is uncertain, but it's likely that Hodgdon put the shifting lever in the wrong gear, for when he stepped on the accelerator the car leaped forward and plunged over the steep bank. Van Dyke either jumped or was thrown from the vehicle onto the ledges, while the driver stayed with the car and was pinned underneath it on the rocks below.

Hodgdon died almost immediately and the injured Van Dyke was rushed to the Farren Hospital, insisting when conscious that his chauffeur was not to blame for the accident. Van Dyke, who was just 64 at the time, later died in the hospital, too. Never having married, he was survived only by his two brothers, Thomas and Philo.

The curious and the morbid gather around the remains of the fatal crash that killed George Van Dyke and his chauffeur. Courtesy of Peter Miller.

The 1909 log drive, meanwhile, continued on, despite Van Dyke's unfortunate accident. The drive finally reached Mount Tom on September 9, some 145 days after it started at Bloomfield in northern Vermont.

Despite the passing of its indomitable chief, the Connecticut Valley Lumber Co. continued its dominance of New England's largest river. Thomas and Philo Van Dyke carried on, placing young George Van Dyke, nephew of the deceased timber baron, in charge of the 1910 log drive. The Connecticut River must have flowed with a good pitch that year, because the log drive containing 42.4 million board feet arrived at Mount Tom the first week in July, only 96 days after its start. That, however, was the only year that young Van Dyke was ever heard from on the river.

The log drive of 1911 was bossed by Donald J. McDonald, the CVL Co.'s northern manager, along with Tom Van Dyke. Charles Hadley of North Stratford was clerk of the drive and Phonse Roby served as the walking boss. By then, J. J. Phelan, a Boston financier, was serving as president of the company and George Lewis as treasurer. The drive that year had 44.5 million feet of CVL Co. logs, plus another 15 million feet of pulpwood logs. It was the largest drive ever, or so it was proclaimed. But that claim had also been made in other years and estimates always varied. Even the contemporary newspapers couldn't agree on the size of the drive. At any rate, the older rivermen said the 1911 drive might never be topped as it most certainly was larger than the average-sized log drive of 50 to 55 million board feet.

The boss of the drive, his assistants, and several clerks, gather outside the CVL Co. headquarters tent during the 1912 drive. The two clerks are Charles Hadley, in the center, and Harry Curtis, left, who, it was said, could speak seven languages. Collection of A. Janion.

Within a year or two, the Van Dyke dynasty came to an end when in June 1913 it was announced that the company had been sold to a syndicate composed of two Boston corporations, Stone and Webster, plus Hornblower and Weeks. The principal attraction for the new owners was not the timber, which was valued at $2.9 million, or the sawmill activities; it was the package of 330,000 acres of land with the accompanying water rights, which were now becoming valuable for power generation.

Howard Philbrook, executive vice president of Stone and Webster, was given control of the Connecticut Valley Lumber Co. His cousin, Earl Philbrook, became the CVL northern manager for the owners; George Monnell was named the company forester and

The Unscrupulous Clerk

In his memoirs, riverman Fleetwood Pride related the following story about a clerk he had never forgotten.

"The following spring I went back on the Connecticut River again for the Van Dyke drive. That year the timekeeper was a Johnny Holmes, and he followed the men a great deal at their work. If he saw a man standing idle, he'd dock him a part of the day. It was said he got a percentage of what he saved the company in that way. Once near the town of Colebrook he saw a scarecrow that seemed to be leaning on a pole. Johnny out with his book put down somebody's name and called out to the scarecrow, 'I know you. You was here last year.' I suppose some poor fellow lost a couple of dollars and probably Johnny got one of them. Later we'd call out to Johnny, 'I'm a scarecrow.' Someone else would chime in, 'I'll be damned if you don't look like it.' He was a little old wrinkled up fella."

Orton Newhall the resident engineer. Monnell resigned in 1924 and was replaced by John Locke.

A number of these professional men came on board, but as the old-timers lamented, it wasn't like the good old days. The CVL Co. no longer existed as a tough, traditional New England logging concern motivated to stand on its own feet financially. Part of the assemblage of old-time woods bosses was still there, but the management now reflected a much softer and urbane character.

Stone and Webster Co. enjoyed their company retreat in the deep woods of northern New Hampshire, their "northern office," and made frequent recreational junkets to the Connecticut Lakes. Old Camp Idlewild on Second Connecticut Lake was rehabili-

The drive boss, who moved around quickly in a buggy, is photographed below Turners Falls dam in this undated photograph. Courtesy of Montague Historical Society.

CVL Co. officials are pictured here at the company headquarters at First Connecticut Lake. Identified are James Garvey and Pat Shea, the first two on the left; General Supt. Hancock, the fourth from left; head bookkeeper Gene Andrews, fifth from left.

tated, a proper access road was put in, and a new motor boat was provided, all for the comfort of company officials.

THE 1914 LOG DRIVE

The spring and summer of 1914 promised to bring about some momentous events for recording in history as the Connecticut River Lumber Co. prepared for what was again promised to be the largest log drive ever held on the Connecticut. At the same, rumors persisted that the drive might very well be the final one.

The winter logging season of 1913–14 had proven to be a busy one for CVL as the company filled the camps along Paul Stream and along the East Branch of the Nulhegan River with 1500 men and 750 horses. By winter's end there were 25 million board feet of 30- to 40-foot-long spruce logs piled on the frozen streambed of Paul Stream, stocked solid on the ice from Brown's Mill to Bull's Throat and from Walker Dam to Granby Branch and a half-mile up the Branch. Twenty million feet of logs were also stacked along the East Branch of the Nulhegan, with about half that volume below Dam No. 3. The Paul Stream harvest logged out the last remaining old-growth spruce in that watershed. Additional cutting along the Connecticut River about four miles above North Stratford pushed the total volume to over 60 million board feet of CVL logs waiting for what would hopefully be an early spring break-up.

Winter-cut spruce logs are piled high on the ice covering Paul Stream above Walker Dam in the spring of 1914. By this time, April 10, the ice was just beginning to show indications of melting in the ponded area. Approximately 25 million board feet of logs were piled on Paul Stream waiting for release. Stone and Webster Report, 1914.

Brown's Mill camp on Paul Stream in Essex County, Vermont, was used by the CVL in 1914 as living quarters for many of the rivermen, and for summer pasturage of the company's farm animals. The camp was formerly the site of a water-powered sawmill settlement. Stone and Webster report, 1914.

And that wasn't the end of the log flow planned for that active year. The CVL drive was to be followed by the International Paper Co. drive of 15 million board feet of pulpwood logs headed for Bellows Falls and another 3 million feet of pulpwood logs waiting behind a boom at Gilman, Vermont. Thus a possible record flow of logs was gathered and waiting; an early ice-out and a moderate flow of water in the Connecticut River would portend a successful log drive.

Such was not to be the case, however. As mid-March arrived the ground began to thaw, and the loggers could be seen strolling in small groups out from the camps deep in the woods. Every available barn in Stratford and Bloomfield was said to be put into shape for the "springing out" of the horses. The wait for the ice to break was usually a short one.

The weather, though, did not cooperate; the temperature continued cold through to mid-April, and the ice in the rivers and streams held tight. Up at Brown's Mill Camp on Paul Stream, 50 impatient rivermen were holed up, anxiously awaiting the breakup, and no doubt grumbling because they would receive no pay until the log drive began. They weren't happy either about being charged a dollar per day for board. They were threatening to quit, and a few did end up heading down the woods road.

There were another 50 men encamped at Bull's Throat, and yet another 50 up at Granby Camp, all making time until action began on the Connecticut River. Drive boss Donald McDonald arranged for a few more men to be shipped in from the employment agencies in Portland and Boston. Once the drive began the men would be paid $3.25 a day, with the better workers getting as much as $3.50 a day.

Finally, on the 16th of April, the weather began to moderate some, and the river rose a couple of inches. Then, two days later,

Early on during the 1914 drive, a small jam occurred a half-mile above the mouth of Paul Stream. A crew of rivermen from Brown's Mill camp were then called in work the logs free. It took them most of the day, but by 3 p.m. on April 14, the rivermen had freed the jammed logs. Stone and Webster report, 1914.

warm weather moved in, and a few logs began to move in some open water. While a few small jams plagued the initial stages of the drive, more serious troubles arrived on the 19th, when the river rose 16 inches in a day, thrusting the moving mass of logs together and causing occasional jamming on gravel bars. At the same time, the swift current packed the logs tighter and tighter.

The worst jam was located two and a half miles below North Stratford where the Grand Trunk and Maine Central Railroads pass alongside the river. McDonald brought down more than a hundred men from the Paul Stream camps, putting them to work picking apart the jam log by log, but then decided to wait until the river level dropped before loosening the jam for good. His concern was the potential danger to the Diamond Bridge down below if the logs went down in a single mass during high water.

The river level continued to rise as the temperature continued upward, and the crisis worsened as additional logs came tumbling down the river to add to the jam, now totaling about 10 million feet of logs wedged tightly together. As the logs piled up against the railroad embankment the river flow was blocked and the current was diverted over the banks, washing out various sections of the railroad bed. The Grand Trunk had to suspend railroad serv-

On April 18, 1914, a rapidly rising river sent a large mass of logs down the Connecticut and resulted in a sizable jam on a gravel bar about 2½ miles below North Stratford. The groaning, jackstrawed logs were backed up for about a half-mile. The tracks of the Grand Trunk and Maine Central Railways, pictured here on the east side of the river, were washed out in a number of locations by the backed-up waters. Stone and Webster Report, 1914.

Food preparation is under way on the "Mary Ann" in this photo from 1912. Collection of A. Janion.

ice on the 20th, and the nearby Maine Central Railroad discontinued rail traffic a day later, on April 21.

Fortunately for the rivermen the temperature began to drop again on the 21st. This resulted in a corresponding drop in the river level, which in turn averted a potential long-term crisis. The irony of the situation now, however, was that the river was suddenly too low for good river driving. The whole situation was complicated by the fact that there hadn't been a heavy snowfall that winter, and the sudden thaw had taken *too* much of the water in the initial melt.

The crews up along Paul Stream were having a difficult time, too, first because of the high water and then because of an insufficient supply of water needed to move the logs down the narrow stream bed, a process which usually took three to four weeks.

Digging apart the main jam in the Connecticut kept a large crew of men busy for a week as they pushed and pulled out the tightly packed logs. Much of the jam now sat above the level of the receding river. Two weeks had been lost due to the adverse conditions, but eventually the log drive was finally under way, albeit with a diminished flow of water in the river.

The Connecticut River remained unusually low that spring and summer, causing extra toil for the rivermen moving their logs southward. The rocky stretch through Fifteen Mile Falls was the usual tough barrier with its many exposed rock formations. Over 200 men went down on the main drive that year, while another 70 or so were sent down the river in advance. Still another large contingent of men followed up at the rear of the drive, cleaning up the meadows and coaxing the stragglers back into the water.

Cook Bill Loveland and helpers display the items to be served to the rivermen at a mid-morning meal. Note the numbering of the sections of the floating wanigan. This made for easy dismantling and assembly when the wanigan had to be taken out of the river to avoid forbidding falls ahead. At the end of the drive, the Mary Ann was dismantled and returned north by railroad.

The troublesome CVL drive of 1914 arrived at the Mount Tom sawmill the first week in October after 173 days en route down the river. The logs were welcomed at the sawmill because the mill had been without any logs through the dry months of July and August, the previous year's supply having been used up months before.

The river was so low in 1914 that the CVL was unable to get the

The cookee had the chore of peeling potatoes on the "Mary Ann." Collection of A. Janion.

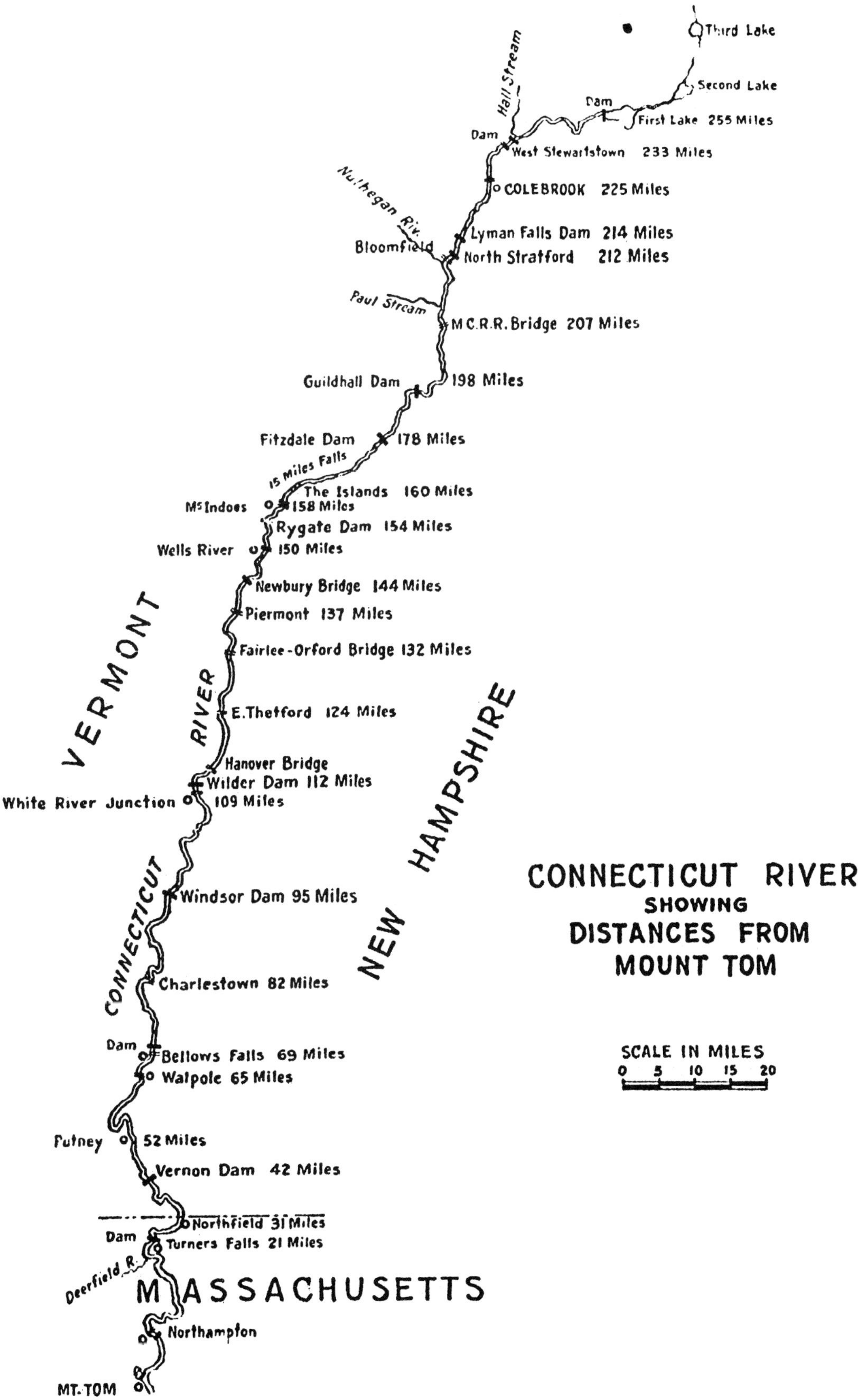
Third Lake
Second Lake
Dam
First Lake 255 Miles
Hall Stream
Dam
West Stewartstown 233 Miles
COLEBROOK 225 Miles
Nulhegan Riv.
Lyman Falls Dam 214 Miles
Bloomfield
North Stratford 212 Miles
Paul Stream
M.C.R.R. Bridge 207 Miles
Guildhall Dam
198 Miles
Fitzdale Dam
178 Miles
15 Miles Falls
The Islands 160 Miles
McIndoes
158 Miles
Rygate Dam 154 Miles
Wells River
150 Miles
Newbury Bridge 144 Miles
Piermont 137 Miles
Fairlee-Orford Bridge 132 Miles
VERMONT
RIVER
E. Thetford 124 Miles
Hanover Bridge
Wilder Dam 112 Miles
White River Junction
109 Miles
NEW HAMPSHIRE
CONNECTICUT
Windsor Dam 95 Miles
CONNECTICUT RIVER
SHOWING
DISTANCES FROM
MOUNT TOM
Charlestown 82 Miles
Dam
Bellows Falls 69 Miles
Walpole 65 Miles
SCALE IN MILES
0 5 10 15 20
Putney
52 Miles
Vernon Dam 42 Miles
Northfield 31 Miles
Dam
Turners Falls 21 Miles
Deerfield R.
MASSACHUSETTS
Northampton
MT. TOM

entire log drive into the lagoon on the river's oxbow at Mount Tom. Thus about half the logs had to be piled up on the bank, hauled there by a crew of 120 men, five teams of horses, and a handful of steam donkey engines. The delivered logs were valued that year at $21 per thousand board feet.

The 1914 drive was an exhausting and expensive effort for the CVL Company. But that's the way things turned out some years, as circumstances often beyond the river drivers' control usually dictated how events would unfold from one end of the drive to the other. For the most part, the rivermen were at the mercy of the fickle forces of nature.

For example, the arrival of the head end of the log drive at Mount Tom could occur any time between mid-July and mid-October, and the length of time on the river could be anywhere from 75 to 190 days. The variance depended on the pitch of the water flow that year and the date when the drive got started on the north end. The average time for the drives was about four and a

Tom Cozzie

The year 1914 is also noted for the appearance in the North Country of Russian immigrant Tom Cozzie, soon to become well known as a titan in the woods. Born September 15, 1894, near Kiev in the Ukraine, Tom had heard about the United States from his father and uncles who had previously been to this country and actually worked for the CVL for a year or so. Borrowing some money from his father, who had returned home, Tom came to the U.S. and worked for six months on a trolley line being built between Berlin and Gorham, New Hampshire.

In July, Tom arrived at the West Stewartstown railroad station with three of his countrymen, none of whom spoke a word of English. He was put to work as a swamper at a camp near Ferdinand Bog, Paul Stream, and soon worked his way up to be a chopper. At the end of the logging season Tom was kept on as a handyman and cook for a survey crew headed by Win Schoppe. From that humble beginning Tom Cozzie worked his way up to become the woods superintendent for St. Regis Company, the later owner of the vast Van Dyke timberlands.

Tom Coozie displays his playfulness at a camp on the East Branch of the Dead Diamond River in 1926.

half months, with the distance from Bloomfield to the Mount Tom sawmill being about 218 miles.

The start time of the drives also varied from year to year. They were known to have begun as early as March 30 and as late as May 5, depending primarily on when the ice broke up. Most years, however, the rivermen were into the icy water by mid-April.

As difficult as the 1914 drive was, making it even tougher for CVL employees was the persistent rumor that the 1914 log drive was to be the last of the CVL drives. This rumor, which had circulated through the camps all winter long, was similar to rumors heard through the woods the previous two or three years. All company officials would say was that timber harvesting would continue indefinitely. The loggers shuddered to think that the era of their cherished occupation could possibly come to an end.

Despite the rumors, the future did look optimistic for the woodsmen by late summer of 1914, even after the log drive had concluded, for the company already had 500 men in the woods and 210 horses fattening up in pastures. By the height of the 1914–15 logging season in mid-winter, the CVL Co. had 1500 men and 750 horses busily stacking spruce logs streamside in preparation for the spring drive of 1915. In all they had 18 camps operating in the towns of Lewis, Ferdinand, Maidstone, and Granby, Vermont, and in Columbia, New Hampshire.

As the winter logging season wound down, the next rumor persisted; 1915 would be the last year for a log drive down the Connecticut River, it was said. The thought of life without the annual log drive was an unpleasant one for many of the veteran rivermen.

Then the rumor became official, confirmed. The year 1915 *would* witness the last of the great Connecticut River log drives as CVL Co. was giving up the momentous endeavors. At the news, every woodsman who had ever worked on the river scrambled to be on hand that spring; they all wanted to be among the ones chosen to go down the Connecticut in 1915. What a cherished honor it would be to go down the river on the finale of the great log drive era.

13

The Last Log Drive

THE LOG DRIVE of 1915 was a big one. In size, yes, but especially in the achievements and memories of rivermen all across northern New England. In the late winter the rumor that had been circulating among the logging camps and among the logging fraternity throughout Maine and elsewhere was confirmed as factual. This would be the final long log drive that the CVL Co. would send down the river. The sad news became the topic of chatter in camps and saloons all over the North Country, for there were those who had maintained that there would surely be log drives on the Connecticut River until the end of time.

It is certainly understandable why so many rivermen, young and old, traveled to the North Country that winter. Log drivers were passionate about their chosen and dangerous occupation; each and every one wanted a piece of the action that year. That is, if they were to be among the privileged chosen ones.

The timber cutting operations that winter had been concentrated on the remaining timber stands around the Connecticut Lakes in far northern New Hampshire. As mentioned previously, the once vast old-growth timber resources of Paul Stream and the Nulhegan River had almost reached depletion during the previous year's winter harvest.

There were well over 500 men that went down the Connecticut on that final drive. Included amongst the crew were at least two old-timers who had been part of the crew going down on that very first loose log drive held by the company in 1869. Al Patrick was just a youngster from Maine back then, when he hired on to the drive under drive boss Charles Weeks. Afterward he went back to Maine for a couple of years, returned to the Connecticut River in the early 1870s, and was part of most drives from that date onward. Also on hand for the 1915 drive was Rube Leonard of Colebrook, who claimed to have worked with every drive since 1869.

Donald J. McDonald, the CVL northern manager, was again boss of the drive. The competent Phonse Roby bossed the head of the drive and in charge of the rear was the renowned Win Schoppe, another veteran woodsman. Dan Bosse, the celebrated log driver from over on the Androscoggin River, also showed up that spring. It is said that even some of the old Bangor Tigers came over from the Penobscot River in Maine.

Elsewhere that spring, Bill Bacon was given charge of the horse rafts on the floating rear, while veteran riverman Tom Cozzie was among the crew sent up to swamp out the small river between Third and Second Lakes in order to retrieve some of the logs previously left in the lake by Van Dyke's crew.

There wasn't much time for the sharing of tales of the old days, however, for ice went out in late March that year, and logs stacked streamside up along the Second Connecticut Lake began to roll into the cold lake water.

Down across First Lake and over the sluiceway at the First Lake Dam rolled as much as 50 million board feet of logs. Tumbling on down the turbulent waters of the Connecticut River, the mass was joined by more logs from Dead Water Stream and Perry Stream. Before the logs had cleared the area, two rivermen had tragically drowned at Perry Falls, but there was no stopping the drive at that point.

Though the onward movement was held up for a short time by the boom at North Stratford, where the logs accumulated together into a solid bunch resulting in a jam, the drive officially got underway on April 9, soon thereafter logs from the Nulhegan River, Paul Stream and other tributaries were added to the mix. On down the river they bumped and jostled.

Upon arrival at Fitzdale (Gilman) there was a temporary impasse, as reported by historian Robert Pike. The management of the paper mill there refused permission for the rivermen to sluice their logs over the dam. Unlike the days when George Van Dyke ran the company and its drives, the new owners of the CVL Co. lacked the tenacity and bravado of the old-time river bosses, and thus manager Donald McDonald ordered the drive be halted until the dispute could be settled.

But tough old Phonse Roby was from the "old school" of Van Dyke, and he did not want to wait one minute as he envisioned the possibility of the water level dropping in the river's headwaters. He said nothing to McDonald but was soon seen walking into the main office of the paper company, his spiked boots gouging little holes in the polished hardwood floors. His words to the mill manager

When the logs started moving again on this logjam between Monroe and Barnet in 1915, the men must have had a ways to scramble for safety.

Observers watch the river crew at work in 1915, just above McIndoe Falls. From the left: Philo Van Dyke (George's brother), unidentified river boss, Murray Way, Horace Gleason, and Darwin Nichols.

were something to the effect that they had best open those sluices or God have mercy on the owner when Roby turns loose the rivermen, because there might be nothing left of the dam. Roby meant what he said, his antagonist knew it, and the drive went through without delay.

Onward the drive continued, first through the Fifteen Mile Falls and Mulliken's Pitch, and then on by the Twenty-Seven Islands and into the troublesome Woodsville Narrows. By then, it is claimed, the volume of the drive had reached 65 million board feet of timber. Meanwhile, ahead of the CVL drive, was the International Paper Co. drive of 15,000 cords, which had come down the White River and arrived at White River Junction on May 21.

As there had been heavy rains that spring, the Connecticut River had risen much higher than usual. The swift water pushed the CVL logs rapidly along, but at the same time created problems for the rivermen trying to control the assemblage of logs. A boom had been put across the river above Bellows Falls to concentrate the logs and govern the flow over the dam. The hope was that it would hold.

It's feeding time for the horses grouped together on a 16-horse raft.

The drive arrived at Bellows Falls on July 25, with the Connecticut River still running at fever pitch; the boom, it turned out, couldn't hold them all back. Many broke through and had to be stopped by another boom strung across at North Walpole. Boat owners now worried that their crafts still in the river would be crushed.

With the rear of the drive still way up at Fifteen Mile Falls, word was sent back upriver to string the boom across at Wilder and hold the logs back there for a time. The floating wanigan, Mary Ann, was also assembled and put in the water at Woodsville for the last time.

The river was high enough over the Bellows Falls dam that there was no need during that summer of 1915 to shut off the water through the industrial canal. There were no complaints from the mill owners nor harassment from idle mill workers.

The larger horse rafts would need about a dozen men to pole them down the river. The rafts were put in at McIndoe Falls or at Wells River, and enough would be used to carry about 40 horses. The fourth man from the left is Dolph LeClaire, who lived in Lancaster in later years and shared some of his experiences with the author. Collection of A. Janion.

Driving logs down a river was, by necessity, very labor intensive and it didn't take long for costs to escalate. Here, the rivermen are attempting to gain some open water in order to free this wing jam in 1912. The top log has a boom chain attached and the log under it had previously been blown free with dynamite. Collection of A. Janion.

The logs flowed steadily over the dam and down through the ledges, but the rivermen had to work hard at the Falls. The day crew was up for a 4:30 breakfast at the camp on the New Hampshire side, a quarter-mile below the old Tucker toll bridge. On the river by 5 A.M., they ate again at 9 A.M. and 2 P.M., the food carried to them by a wagon. Quitting at 7 P.M., they were then replaced by a night crew, which worked under lights mounted on the bridge until 5 A.M., pushing the logs back into the current both above and below the dam.

By August 8 the drive had cleared Bellows Falls, except for a few stragglers, and the men continued to move on down the river. But more troubles with log booms were still ahead.

Arriving at Turners Falls, the logs massed together behind the boom that was customarily stretched across the river above the dam in order to sort out the logs that belonged to the Turners Falls Lumber Co. But again the pressure of the high water current was too much; the boom parted and eight million feet of logs were released, soon to jam against the Hadley bridge. The jam extended for 2,000 feet up the river and was wedged in tightly against the piers. Forty men toiled for three days to work the logs through a 200-foot wide passage between two bridge piers.

The arrival of the logs at the mill in early September and the herding of them together into a large boom enclosure were occasions of considerable fanfare. The many onlookers along the riverbanks watched as the rivermen displayed their skills atop the logs for the last time. Old-timers such as Al Patrick and Rube Leonard shared what had to be a sad closure to an exciting era along the flowing water of the Connecticut River. Both Patrick and Leonard had been members of the river crew that brought down that first log drive for the company in 1869. Though there was no further announcement from the front office of Stone and Webster, everyone on hand assumed the foregone conclusion that the glory days on the Connecticut River were history.

Even with the conclusion of the 1915 log drive, CVL's troubles were not over, for labor discontentment was escalating in the big sawmill at Mount Tom. On January 10, 1916, the sixty yard workers there went on strike, demanding $2 per day pay instead of $1.65. This forced the mill to temporarily shut down the following day, putting two hundred men out of work. Other strikes occurred periodically, such as a walkout of the laborers on the lumber platform, which also shut down the mill for a few days.

By late fall of 1916 company officials decided they had seen and had enough, and on November 6 the sawmill shut off the saws permanently. Estimates were that a year would be required to finish up the shipments of sawn lumber in the yard. Before the month of November was gone, though, the announcement came that the Mount Tom Sulphite Pulp Co. would buy the land, plus

A steam-powered paddle wheel scow is used to herd the logs into the holding area at the Mount Tom sawmill. This photo was identified as having been taken at the end of the final log drive in 1915.

the unused logs on hand. The transaction became final a few months later, in April 1917.

Despite the sale of the Mount Tom mill, which was preceded in 1906 by the closing of the CVL Co. sawmill in Holyoke, and the earlier discontinuation of the Connecticut Valley Lumber Co. as a manufacturer of lumber, the company was still by far the largest landowner in the Connecticut River watershed.

THE FACTORS BEHIND THE FINALE

Why did the river drives of long-length logs come to an end? After all, the energy used to move the logs over two hundred miles to the sawmill was in a sense free and available for most anyone to use. And for many generations the Connecticut River had been used as an avenue of commerce, for the transport of logs, lumber, and agricultural products and people.

A change in circumstances came about which had a two-fold motivation—economics and environmental. Log drives had begun to incur increasingly more expense for the CVL and the other companies using the river. Costs by the second decade of the 20th century were considerably greater than those reported for 1887, which even then exceeded $2,000 per day on the drive. And in 1915 the International Paper Company reported it was having problems finding a sufficient number of competent rivermen to handle their pulpwood drive on the White River.

Adding to the increased costs was the decreasing size of the logs, which intensified the unit costs. William H. Gray, a Boston lumber merchant who had connections with the CVL, made note in his communiqués that the spruce logs were steadily becoming smaller. Where it formally took an average of five trees or logs to produce a thousand board feet of lumber, he pointed out that by 1915, fifteen logs were required to produce the same amount of lumber. The CVL had not cut off all of the old growth spruce, but they had apparently removed the best of it in the earlier years and were cutting on the less productive sites in later years.

But the greatest problems affecting the CVL log drive were those of an environmental nature. Logs caused damage to the riverbanks, the meadows, and the bridges, and lawsuits filed against the company had become expensive to defend. Although the lumber companies had traditionally been granted prerogative on the river as a public thoroughfare, the voices of competing users of the river began to be heard in the early 1900s. Public sentiment toward the log drives was now changing.

The agitation had actually begun back in the early 1880s when a man named John French complained that he couldn't get his small passenger steamboat across the Connecticut River at Holyoke, Massachusetts, because of the logs massed there for use at the company mill. He eventually brought suit against the Connecticut River Lumber Co. to prevent them from blocking the river.

And the agitation continued to grow. In July 1908 boat owners in the Bellows Falls area sought an injunction against the CVL to prevent the blocking of the river with logs. On July 6 of that year the rivermen had put their boom across the river up above Bellows Falls, but in doing so had cut off river passage to the many riverside cottages up above. The fact that the logs weren't expected to arrive for at least two weeks after the boom was set into place further aggravated the situation. An injunction was granted to prevent the blocking of the river by the CVL, prompting company executives Thomas Van Dyke and William A. Moore to come down the river in an attempt to negotiate and hopefully prevent an action that would tie up the drive. The CVL was then forced to agree to stretch the boom further upstream, at South Charlestown, above the last riverside cottage, and to hold the entire drive there until all the logs had arrived, after which they would be let loose and hurried over the dam.

During that same summer of 1908, the boat owners on the river between Turners Falls and Holyoke were planning measures to prevent the blocking of the river with logs. Their claim was that they too had rights on use of the river, since the courts had previously declared the Connecticut River to be a navigable river. During the ensuing years there was talk of new state legislation regarding logs on the river, possibly requiring the lumber companies to undertake the costly measure of tying the logs together into raft form once they had entered Massachusetts waters. In light of these new developments, CVL officials quietly feared that their Mount Tom mill would have to be closed down.

The log drives continued, however, and so did the legal problems. In 1912, the residents of Bellows Falls, irritated not only by boating being at a standstill during the height of the summer season but also by the local mills being idle during the period that the water canal was closed off, were calling for new legislation at the state level. A thousand angry men were idled and a $25,000 weekly payroll suspended because of the log drive. In response, businessman Hayes Bigelow filed suit against the CVL because of damage to his business.

The boat owners, seeking to keep the river open for travel, also succeeded in having an injunction served against CVL the following year, while at the same time, on the New Hampshire side of the river, a court in Keene ordered that a channel be kept clear for boats from Putney, Vermont, south to the Vernon (Vermont) dam. Each year, it seemed, it was becoming more difficult for the river drives.

While the CVL Co. had previously given some consideration toward constructing a sawmill nearer the timber source in the North Country, that idea died with the sale of the company to Stone and Webster in 1913. Problems on the river also continued for the new owners, even as they pushed through one of the largest log drives ever in 1914.

In February 1915, engineers working for Stone and Webster began looking over some property north of Bellows Falls with the consideration of pulling the entire drive out of the river at this point. The handling of the logs and installation of booms were becoming so expensive below that point on the river that the company considered jacking the logs out, cutting them into shorter log lengths, and loading them onto railcars for transport to the Mount Tom mill. Thought was even given to building a sawmill or a sulphite pulp mill at this location. In March, however, Howard Philbrook wrote the business manager at Bellows Falls, denying any intent to build a mill above the Falls.

Thus the old traditional Connecticut River log drives were a victim of changing times. The decision to cease the drives in 1915 was basically a matter of rising costs for the company. And one of the major economic factors behind the decision was the legal hassles and restrictions placed on the CVL Co. because of competing uses for the river.

14

Later River Drives of Short Length Pulpwood

ALTHOUGH THE long-log drives had come to an end in 1915, the memorable log drive of that year was not the complete end of industrial wood being transported down the Connecticut River. During most of the next 33 years there were sporadic drives of four-foot-long wood floated down a portion of the river, though never over the distance that the long logs once traveled. Four feet had by that time become the customary length to cut and deliver pulpwood for paper manufacture. Some of these short wood drives were fairly large in volume, though never as immense as the river drives of the CVL.

River driving of short four-foot wood was somewhat different than driving long logs that were 30 to 40 feet in length. The short bolts would twist and tumble over each other, sometimes in mad confusion, but would move along quickly in the current. Short wood would jam, but not as readily and severely as their long-length brethren would do. Obstructions in the river such as boulders would catch a few sticks, possibly requiring a man to "tend out" at the spot with the long pike pole used to push the bolts free, but they would usually move readily. Cleaning the wood off hang-ups with a pike pole could be difficult while having to stand out in the swift cold current. Another man might stand upstream from the worker to break the water current and lessen pressure on the worker.

Pulpwood logged in the spring and early summer would more often than not have had the bark peeled off by the logger, because the bark could easily be removed during that season. The bark

Four-foot rough or unpeeled pulpwood is stacked and dumped along the course of Smith Brook, which normally did not flow enough water to float the wood. By closing the dam below, the area would be flooded and the wood floated down into First Connecticut Lake. Courtesy of Perly Hurd.

always had to be removed, either on the logging job or by a debarker at the mill, before the wood could be utilized in the paper-making process. This "peeled" wood would be relatively dry by the time of the river drive the next spring and would ride high in the water, making less labor for the rivermen. But pulpwood cut in the latter part of the summer and in the winter would still have the tight gripping bark in place on the wood. This unpeeled wood, referred to as "rough" wood, was heavier, rode lower in the water, and tended to hang up easier on a river drive because of the rough surface of the wood.

Floating in a relatively smooth river flow, the four-foot wood would flow steadily, resembling a carpet covering the river surface. But the short pieces had their disadvantages; they would frequently lose their way out of the current into an eddy, or easily

A recently built splash dam is viewed from the upstream side in this undated photograph along a stream in northern New Hampshire. The small gate will be used to pass four-foot pulpwood.

The sluice gate has been opened and the short pulpwood sticks have begun journey from the Connecticut River headwaters in 1918.

Peeled CVL pulpwood filled the Connecticut River at the South Lancaster boom during the 1916 river drive.

Booms sometimes fail in high water. This boom at Cones Holding, above North Stratford, couldn't contain the pulpwood under pressure and lost the wood. Note the upright sticks set in the holes drilled in the boom logs to prevent the four-foot wood from going over the top. The pier in the center of the river for securing the boom has a capstan on it. Courtesy of Paul Glover.

wander out of the river when the water overflowed the banks, or quickly became grounded in shallow riffles.

Driving pulpwood down a small stream required careful control of water release at the dams along the way. Too little water could leave hundreds of four-foot sticks stranded all along both sides of the watercourse. On the other hand, too much water could exhaust the needed supply before completion of the drive. Should that happen, the crew tending the rear might have an exhausting chore before all the wood was down.

The techniques of the rivermen handling four-foot wood were,

Peeled pulpwood gathers behind the boom at Schoppe Dam flowage at Second Connecticut Lake.

therefore, different from those used on the glorious long-log drives that most people think about. For the pulpwood rivermen there were no romantic tales of the daring man who rode the logs down through the rapids because a four-foot bolt would not support a man's weight. There were no exciting accounts of the rivermen who challenged death to break a logjam. There were tragedies, of course, on the pulpwood drives, but even they were not as common.

The Connecticut Valley Lumber Co., now that the large-scale harvest of sawlogs had ceased, turned to pulpwood logging on their vast timberlands. Beginning in the spring of 1916, the company loggers commenced dumping large quantities of four-foot pulpwood into the Connecticut River. Heavy cutting continued up in the Nulhegan River country, and in the area around the Connecticut Lakes. This was during the era of the First World War, and labor to cut and peel four-foot wood was in short supply. For that reason, the company cut the pulpwood in the Connecticut Lakes region into long-log lengths and moved it down Second and First Connecticut Lakes in large booms that were gathered and closed by means of a donkey steam engine. Containing as much as 3,000 cords of wood, the booms were hauled to the lake outlet with a motorboat.

At the outlet of First Connecticut Lake the company set up a slasher mill, consisting of a series of circular saws at four-foot intervals to "slash" or cut the logs up into pulpwood lengths. This unique slasher mill was powered by a turbine installed within the First Lake dam. The turbine also powered the conveyor that car-

The later dam at First Connecticut Lake with a pulpwood slasher mill and chute built on the top.

Pulpwood headed for the International Paper Co. mill at Bellows Falls is removed from the river by conveyor at a North Walpole site across the river.

ried the logs sideways up over the top of the dam where they then slid down against the bank of saws. The four-foot-long pieces then fell on down into the river below.

In 1921 the CVL ceased logging the pulpwood with their own crews and began to contract with various paper manufacturers and logging contractors to cut the timber and drive the pulpwood on down the Connecticut River to different locations along the river, where it would be hauled out by chain conveyor.

International Paper Co. was one of the largest customers during this era of the teens and early 1920s; the short wood was floated down to the paper mills at Wilder and Bellows Falls. Wood for the Parker Young Co. paper mill in Lincoln, New Hampshire, was hauled out of the river at Mt. Orne (South Lancaster) and at McIndoe Falls for loading on trucks and on the railroad. Both the Gilman Paper Co. mill and the Groveton Paper Co. received their pulpwood via the river from CVL lands on Nash Stream. Logging contractors Marshall McDonald and Wilbur Marshall harvested pulpwood in the Nulhegan valley, hauling the wood out of the Nulhegan River at Stone Dam and loading it on the Grand Trunk Railroad for shipment to Brown Company's mill at Berlin, New Hampshire.

The management of its vast timberland holdings was never a strength of Stone and Webster Company, and after thirteen years of ownership of the former CVL land the engineering company opted to get out of the timber harvesting business. In 1927 they sold their entire northern forest holdings—some 400,000 acres—to the St. Regis Paper Company. The purchase of the vast North Country timberland was a large and costly step for St. Regis. At one point, International Paper Co. was reportedly going to join up with

St. Regis in the purchase of the forestlands, but I.P. Co. had to back out of the transaction as it did not have the available cash due to internal problems within the company.

St. Regis then set up a subsidiary corporation, the New Hampshire–Vermont Lumber Co., to own and manage the huge property. Meanwhile, waterpower lands and rights on the Connecticut River were sold to subsidiaries of the New England Power Association System, also in 1926. This opened the way for a new major harnessing of the Connecticut River for power development. The sale of the waterpower lands and rights by St. Regis reportedly recouped a large part of their cost of purchase from Stone and Webster.

With the sale of the CVL land to St. Regis Co., there came about further changes in personnel and in the way the company ran its woodlands operations, to the further dismay of the old-timers.

Large quantities of pulpwood were conveyed out of the river at McIndoe Falls and loaded onto railroad cars. This facility was built in 1919 and closed in 1927.

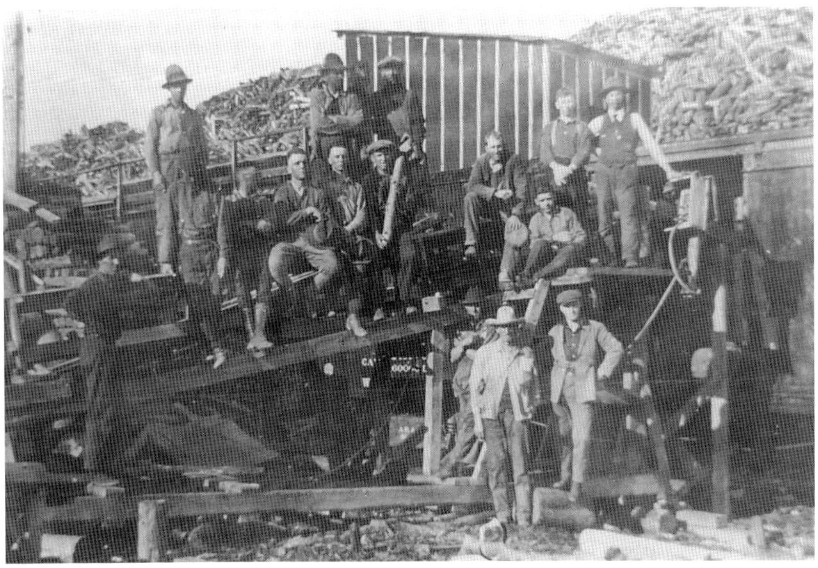

As this photograph from 1923 shows, a conveyor was used to load the pulpwood into railroad boxcars.

In later years, the Connecticut River was used as a log storage location just above the Canaan–West Stewartstown bridge. But the logs went no further than the sawmill on the other side of the river. Note the raft-mounted steam donkey used to stack the logs. Van Dyke Hill, seen in the background, was used to pasture the timber barons' cattle.

Many of the long-familiar and competent faces were removed from the North Country scene. The company appointed John Locke as manager of this northern office, but he was soon let go and replaced by George Kezar.

The short pulpwood drives on the Connecticut River continued under St. Regis ownership, but now the wood was being cut primarily for the St. Regis Co. paper mill in Deferiet, New York. In 1928 the company had Perley Hurd and crew build new piers for booms and a haul-out conveyor in the river at the W. H. Kimball meadow near North Stratford. A pulpwood loading station was built there on a siding of the Grand Trunk Railroad, over which the wood could be shipped to Deferiet. The location was known as Mapleton, and was situated just below the bridge connecting Maidstone, Vermont, and North Stratford.

Perley Hurd had started work on the Connecticut River at age 15 and was praised by his associates as being as nimble on a log as any riverman around. Hurd always said, however, that Dan Bosse was the best man that he ever saw on a log.

In the years following the end of the great long-long drives, the International Paper Co. continued to drive pulpwood down the Connecticut River from their woodland properties in Vermont and New Hampshire. The I.P. Co. had also switched in 1916 to driving four-foot wood down the Connecticut River. In 1919, I.P.

drove 24,000 cords of pulpwood down to the Bellows Falls mill, a full year's consumption. One hundred men were hired that July to remove the wood from the river and stack it riverside on the New Hampshire side of the river at North Walpole.

Pulpwood floated down the Nulhegan River continued to be hauled out and loaded on the railroad at Stone Dam until about 1935. Then there arose a major obstruction to river activity. The completion of Murphy Dam, a large flood control structure on the Connecticut River just above the village of Pittsburg, New Hampshire, put an end to the river driving of any wood from the Connecticut Lakes. Murphy Dam was built with no provision for the passage of logs or pulpwood, despite the earlier assurances given to St. Regis that a sluiceway would be installed.

Floating pulpwood down the Connecticut River began again in 1943 and continued through the years of World War II, with the last wood in and out of the river in 1949. Most of the wood cut in the 1940s was hauled out of the woods by truck and dumped into the river at West Stewartstown and other convenient locations. Wood trucked out during the winter months was landed on a meadow in Canaan, Vermont, opposite the Coos County Farm in West Stewartstown, and thrown in the river the next spring by a crane equipped with an orange-peel bucket.

Almost forty thousand cords of rough spruce and fir pulpwood was driven down the Connecticut during these wartime years to haul-out conveyors located in the northern part of Lancaster, just below the Northumberland town line. The wood was passed through a large drum, debarked at that site, and loaded onto the railroad for transport to Deferiet.

By 1949, more efficient means for transporting pulpwood had

Four-foot pulpwood tended to stray out of the current. In this image from 1948, a clean-up crew is pictured with St. Regis Company wood below Lyman Dam.

been adopted, and all river activity ceased on the Connecticut. The eminent and remarkable log drives, begun about eighty years before, had now come to an end. Thus the great era of log drives on the Connecticut River came to a complete closure.

Epilogue

The legacy left by George Van Dyke was an outstanding estate, one of the largest accumulations of timberland in the northeastern U.S. ever put together by a single individual. For forty-five years his wealth of old-growth timber, covering over 400,000 acres, supported those colossal log drives as 50 to 70 million board feet of spruce and fir sawlogs went down the Connecticut River every year, almost in one solid mass. The harvesting and river-driving activity from his timberland provided employment for literally hundreds of men and supplied the country with the material needed for the construction of thousands of homes and other buildings, many of them still standing.

To get a concept of the enormity of this wood flow of Van Dyke's, compare it with the present-day annual timber cut in the entire state of Vermont. At the time that the old-growth spruce forests were about gone, soon after the end of World War I, the lumbermen turned to harvesting the second growth or next generation of trees. Forests are, of course, renewable, even after the heavy cuts such as those of Van Dyke's loggers. For many years during the early twentieth century the state-wide harvest of spruce and fir timber remained somewhat below the volume of sawlogs which Van Dyke's CVL Co. harvested and drove down the Connecticut River at the turn of the century. And the CVL Co. was harvesting timber *only* on their company-owned property within Essex County of Vermont and Coos County of New Hampshire.

Then in the 1980s, as more of the new forest growth matured and as the market for native spruce lumber again gained strength, the spruce timber cut reached an average of about 45 to 50 million board feet of sawlogs throughout the entire state of Vermont. And by the 1990s, Vermont's annual cut averaged about 72 million board feet of spruce and was then about equal to the CVL Company's largest log drive.

Because of the remarkable reproductive capacity of our woodlands, Vermont is now growing its third and fourth forest growth. The state's total sawlog harvest of all tree species, exclusive of pulpwood, runs 260 to 300 million board feet annually.

George Van Dyke's legacy of CVL Co. timberlands has continued to contribute to the State's economy and to the public benefit for many years. But while one might think or hope that such a large area of contiguous ownership would offer a unique opportunity for long-range forest land management in an area where the population remains sparse even to this day, sadly, that has not been the case.

After the large Van Dyke tract was sold to the St. Regis Company in 1927, the land was managed with applied forestry methods with the hope of achieving a continuous timber harvest. Van Dyke's woodsmen had cut the forest heavily during the many years under CVL ownership, but even so, under the professional management of the St. Regis foresters the land continued to produce large cuts of pulpwood. Road systems were also built throughout the land to replace the now obsolete river drives.

George Kezar was the resident manager for St. Regis during the company's early years of

The stream between First and Second Connecticut Lakes flows peacefully now, hardly indicative of the thousands of spruce logs which once rushed and pounded down over the rough stream bed. The spruce forest is renewable, as noted here.

ownership, joined in 1941 by Fred Cowan, who later served as resident manager from 1950 to 1971. There were even a very few of the old-time woodsmen still around and working for the company in Fred Cowan's day, men such as scalers Rube and Barney Hall, Cleve Dore and Sam Wright, river driver Perley Hurd, and woods superintendent Tom Cozzie.

Unfortunately, the efforts by the St. Regis Company foresters for long range forest management did not assure the future of Van Dyke's legacy, as was originally planned. For Van Dyke's legacy was not an inviolate asset. Times were changing, as they were in much of the nation's economy, and the paper industry was feeling the adjustments. By January 1985 the St. Regis Company was no more, the papermaker having been absorbed into the Champion International Corporation.

The viability of timberland ownership primarily for wood production was also fading during these changing times. As Champion International was to soon realize (if they hadn't already known it), the remaining acreage of Van Dyke's legacy was not a land investment that they'd want to retain for future ownership. With no intention of sustained management or long-term planning, some of the land again received the heavy, liquidation cuts reminiscent of the days of Van Dyke's husky woodsmen. Then, in October 1997, the company put up for sale all of their timberlands, not only the remnant of Van Dyke's legacy but also their timberlands in the Adirondacks of New York.

Van Dyke's legacy eventually became shattered by separate land sales. There were still 133,000 acres remaining in Vermont, and this large chunk became the foundation for what is said to be the largest conservation project ever put together in the history of the State of Vermont. A coalition of conservation organizations and state agencies purchased the area for $26.5 million to forestall any possible future development on the old CVL timberlands.

Controversy soon brewed between the conservation agencies and the sporting public over the permitted use of Van Dyke's North Country woods for certain recreational uses. George Van Dyke would probably be left quite bewildered, though not speechless, over the dispute; the immense timberlands once in private ownership and open to use by the public were now in public ownership with proposals being considered to close part of the land to public use. The remaining acreage in New Hampshire, about 171,500 acres, was purchased by the International Paper Co. with the intention of turning the property over to a land trust. Times were surely changing since the days of the river drives.

So Van Dyke's legacy is gone, melted into a maze of private and public ownership that reveal nothing of the legendary days of the logging camps and the rugged woodsmen who felled the mature timber. The Connecticut River is quiet now, and there are none alive who personally recall the exciting era of log drives. There is only scarce evidence among the river waters to give a hint about the lusty labors of the famed river drivers.

Traces of a few boom piers remain, having survived 90 years or more of ice, high water, and decay. But, included with the many and various toils of those who built the nation, are these endeavors of the colorful rivermen on the Connecticut River log drives.

Glossary

BAG BOOM: Shear boom with slack between points where boom is secured, creating a pocket where logs fill in and form a wing jam to deflect logs.

BALKING: Hauling long mast sticks or other long logs along sled roads.

BARROOM: That section of the logging camp where the men slept.

BATEAU: A specially designed, heavy rowboat for river use, about 36 feet long, with considerable free board, broad of beam, and narrowed to a high peaked and narrow pointed bow and stern.

BEAN HOLE: Hole in ground about four feet across for cooking baked beans in a buried pot.

BELLY ROBBER: Slang for a poor cook.

BIRLING: Rolling a log in the water and staying on the log, a sporting event.

BLOW (*in a dam*): A leak, usually in the planking on the face of the dam.

BOARD FOOT: A section of lumber 12 inches long, 12 inches wide, and one inch thick or one of equivalent volume.

BOB SLED: A single set of sled runners with a cross beam on top upon which the front end of the log is placed while the rear end drags on the ground.

BOOM: A line of logs attached together end to end, usually by chains, and used to contain logs in a pocket, to stop the movement of logs in a river current or to guide floating logs past an eddy or obstruction.

BOX (*in rafting*): A section of a disassembled log raft, usually about 60 feet by 12 feet.

BRANDING AXE: A hammer used to mark ownership on the end of the log.

BREAK UP: The time of year in the spring when ice in rivers and lakes went out and the winter logging roads thawed out.

BRIDLE CHAIN: A chain that is wrapped around a sled runner to act as a brake.

BULL PEN: Bunk house at a logging camp. Also called ram pen.

BULL PUNCHER: Oxen driver.

CALKS (or corks): Small spikes, ¼ inch by ½ inch, set in the soles of the boots to eliminate slipping on log surfaces; also referred to cleats used on horseshoes.

CHOPPER: The member of the logging crew who fells the trees and cuts them into log lengths.

COLLYWOBBLES: Dysentery or diarrhea.

COOKEE: The camp cook's helper or assistant.

CRUISER: A man who estimates the volume of standing timber.

DEACON SEAT: The long seat found in the bunkhouse, made from half of a split log, flat side up.

DINGLE: A shed-like structure used to store food or wood at the logging camp.

DIVISION (*in log rafts*): Full sized log raft, usually six boxes joined together, often 180 feet by 24 feet in size.

DOG: An S-shaped piece of metal attached to the end of the skid chain, and was driven into the log to grip it.

DOG DRIVER: Man who pounded the end of the metal dog into the log with a maul.

DOG TRIPPER: Man who released the dog from the log, striking it on the front end to free it.

DOG WARP: A long rope with a hook on the end, used behind horses to pull out a log wedged into a logjam.

DOLING: Closing the opening in a boom to hold back the log drive.

DRAY: A single log sled with one bunk. Synonymous with bob sled.

DRIVING PITCH: High water suitable for driving logs down the river.

EDDY: A current of water turning around in a direction contrary to the main stream, usually circularly. Frequently found below dams or in river coves.

FIN BOOM: A boom of timbers or logs with a series of collapsible rudders on the lower side which force the boom out into the current.

FLOATING REAR: The horses and supplies on rafts with crew at rear of drive.

FLUSHING: Breaking up the ice in a stream by means of a sudden release of a large quantity of ponded water; also the movement of logs out of a steep, rough stream channel.

GUT HAMMER: A large steel ring or triangle to pound on while calling the men to a meal.

HAULING (*of logs in river*): Logs in jam beginning to move on their own under pressure of river current.

HAY HILL: A steep slope on the sled road which required hay or sand scattered on it to act as a brake for the sled runners.

HEAD: The front end of a sudden rise of water resulting from opening the gate of a dam.

HEADWORKS: A stationary windlass, operated by human, horse or steam power, for pulling a boom of logs across a lake.

HEAVES: A lung sickness in horses.

HIGH BANKER: An observer at the river drive.

HOLDING GROUND: The location in a lake or river where logs were held in a boom until time to release.

HORSE DAM: Logs built up in cob-house fashion in the river to deflect logs away from dead water.

HORSED UP: Logs piled up in jackstraw fashion under pressure of a strong current.

HOVEL: Barn used to house the horses.

INK SLINGER: The camp clerk.

JAM PIER: Pier set in river above the booms to stop motion of logs in the river before sorting.

JILL POKE: On the river drive, a log stuck in a stationary position that may cause jam; also the key log in a jam.

KEY LOG: The log holding back a jam of logs behind it.

LANDING: Location where the logs are gathered or piled for loading onto sled or dumping into river.

LETTING DAYLIGHT INTO THE SWAMP: Cutting the old-growth timber.

LOG RULE: A measurement used to determine the estimated quantity of board feet that can be sawn from a log.

LONG LOG DRIVE: River drive of long-length logs, frequently 20 to 40 feet in length, with no four-foot long pulpwood.

LOOSE LOG DRIVE: River drive of logs floating unattached in any raft form.

MAST LOG: An exceptionally large and long white pine log used for a mast or spar in shipbuilding.

MUZZLE LOADERS: Older style bunks in the logging camps that required men to crawl into them from the foot of the bed.

NOSE BAG: The lunch bucket.

PEAVEY: A stout wooden handle about 5 to 6 feet long with a strong pointed steel spike at the end and an adjustable steel hook. Used for prying and turning logs.

PICKING UP THE REAR: Cleaning up the stranded logs at the rear end of the log drive.

PIKE POLE: A long slim wooden pole, 12 to 20 feet long, with a steel point on the end, often twisted to make it easier to grip the log. Used for pushing, pulling and guiding logs in the water.

PLUG BOOM: Boom made up of logs fastened together by ropes passed through holes bored in the ends of the logs and wedged into the holes with wooden plugs.

POWDER MONKEY: The man delegated to set the dynamite charge when freeing a log jam.

RAISING THEM (*with oxen*): Having all of the oxen on a heavy load move together to start the load.

READING THE SHIRT: Riverman turning his shirt inside out and picking off the cooties.

REAR: The tail end of a log drive with the many stranded logs.

RIVER HOG: A river driver.

ROAD MONKEY: Man on the logging job with the responsibility of keeping sled roads in proper condition.

ROLLED: Man robbed while on a drunk in town.

ROLL DAM: See "Splash dam."

ROLLWAY: Similar to a landing. A location where logs are rolled off a sled into a pile for reloading onto a sled or rolling into the river.

RUN DAM: A wall of logs built against a stream bank on a bend in the stream to keep the logs in the current.

RUNNING THE SIEVE: Selecting rivermen to be discharged as the river drive proceeds and requires fewer men.

SACKING THE REAR: Same as "picking up the rear" (*above*).

SCALING: Measuring the board feet or cord content of trees or logs by a man referred to as the scaler.

SCULLING: Propelling a raft or boat through the water by working a short oar from side to side at the stern.

SHEAR BOOM: A boom of logs strung out to keep floating logs away from eddies or obstructions.

SKIDDING: Hauling logs along the ground without the use of wheels or runners.

SKID ROAD: A cleared path over which horses pulled the logs.

Skidway: Elevated cribwork from which logs were loaded onto large sleds.

Slave market: The employment office for woods help.

Sled tender: The woods worker who assists the teamster, pulling the sled around into position, loading and binding the logs, and performing other duties.

Sluiced: A team of horses upset or forced off the sled road on a slope when unable to hold back the heavy load of logs. Injuries and death to the horses were not uncommon.

Snatch block: A pulley used to change direction of the pull of a cable or rope, usually hung on a tree.

Snubbing: The use of cable or rope to regulate the descent of loaded sleds on steep slopes.

Sorting gap: An opening in the main boom through which logs are passed individually for inspection of identifying marks and sorted by ownership.

Splash dam: A dam built across the width of the stream to raise the water level behind it in order to cover over rocky rapids. During a river drive the logs floating down river would tumble and roll over the top of the dam.

Squeak heel: An affliction of sore feet caused by wearing boots that were constantly wet.

Stag: To cut off trouser legs so that they hung just to the boot tops, leaving no cuff to snag on brush or log stubs.

Straw boss: The foreman.

Tailing: Attaching oxen on the rear end of a large log to act as a brake on steep slopes.

Teamster: The man who handles the horses.

Tend out: Stationary rivermen placed at possible problem locations on the river to keep the logs moving.

Tote team: Horses with wagon or sled used to haul supplies to the logging camp.

Turnout: A short sled road parallel with the main sled road onto which the returning empty sleds could go to avoid the loaded sleds which had precedence.

Turkey: A sack containing all of the personal belongings of a lumberjack or riverman.

Twitching: Dragging a log along the ground.

Two-sled: A sled composed of two sets of sled runners, one under each end of the logs, and connected by a set of cross chains.

Walking boss: The superintendent in charge of a number of logging camps, often walking from camp to camp.

Wanigan (*has several spellings*): The company store at a logging camp; the outfit on a river drive consisting of the wagons, tents, blankets, cooking equipment, etc.; the location where the rivermen make camp for the night.

Wharfing: Building a log skidding road along a side hill by building up with log material.

White water: Swift and turbulent current.

White water man: A river driver who was a good riverman in fast water.

Wing jam: A crib work or assemblage of logs built to deflect logs away from obstructions and back into the current.

Wing up: Small wing jams being allowed to form along sides of river to keep logs flowing in current.

Woods boss: The man in charge of the entire logging operation.

Yarding: The assembling of logs in one place for loading.

Bibliography

BOOKS AND ARTICLES

Bacon, Edwin Monroe. *The Connecticut River.* New York: G. P. Putnam & Sons, 1906.

Bearse, Ray. *George Van Dyke and the Connecticut Valley Lumber Company.* Unpublished manuscript.

Bittinger, John Quincy. *History of Haverhill, N.H.* 1888.

Blaisdell, Katherine. *Over the River and Through the Years—Book Three.* North Haverhill, N.H., 1981.

Brown, William R. *Our Forest Heritage.* Concord, N.H.: New Hampshire Historical, 1958.

Childs, *Hanover. A Bicentennial Book.* 1961.

Cowan, Fred W. "Logging the Connecticut Headwaters." *The Northeastern Logger,* April 1955.

Farrar's Illustrated Guide Book. 1885.

Frizzell, Martha. *History of Walpole, N.H.* 1963.

Frizzell, Martha. *Second History of Charlestown, N.H.* 1955.

Gibson, Alexander D. "Log Driving on the Connecticut." *The Granite Monthly,* October 1923, pp. 497–501.

Hayes, Lyman S. *History of Rockingham, Vt.* 1907.

Holbrook, Stewart H. "The Drive Is Coming Down." *American Forests,* April 1930.

Holbrook, Stewart H. *Holy Old Mackinaw.* New York: Macmillan Co., 1938.

Holbrook, Stewart H. *Yankee Loggers.* New York: International Paper Co., 1961.

Hunt, Orrin Roberts. "A Winter in a Logging Camp." *The Granite Monthly,* February 1896.

Johnson, Frances Ann. *The History of Monroe, 1761–1954.* 1955.

Lovell, Francis & Leverett. *History of Rockingham.* 1958.

Meader, Katherine C. "The Last Log Drive." *The Granite Monthly,* October 1919, pp. 444–446.

Report of the New Hampshire Forestry Commission. Concord, N.H., 1890.

Pike, Robert E. *Spiked Boots.* St. Johnsbury, Vt.: Cowles Press, 1959.

Pike, Robert E. "River Driving in New Hampshire." *New Hampshire Profiles,* May 1967.

Pike, Robert E. *Tall Trees and Tough Men.* New York: W. W. Norton, 1967.

Poole, Ernest. *The Great White Hills of New Hampshire.* Garden City, N.Y.: Doubleday & Company, Inc., 1946.

Pride, Fleetwood. "Fleetwood Pride 1864–1960." *Northeast Folklore,* Vol. IX, 1967.

Roby, Joe Jr. *Family Sketch.* Unpublished manuscript.

Sherman, Leonard Dewitt. "Jack Straws: The Riverman's Game." *Recreation,* March 1910.

Smith, David C. *A History of Lumbering in Maine: 1861–1960.* Orono, Maine: University of Maine Studies, 1972.

Sumner, David H. Family papers, Vermont Historical Society.

Tarrant, Isabel. *Dam It All.* Self-published, 1973.

Underwood, Marsh. *The Log of a Logger.* Self-published, 1938.

Van Dusen, Albert Edward. *Connecticut.* New York: Random House, 1961.

PERIODICALS

Bellows Falls Times, Bellows Falls, Vt.

Brattleboro Daily Reformer, Brattleboro, Vt.

Burlington Free Press, Burlington, Vt.

Coos County Democrat, Lancaster, N.H.

Daily Free Press

Daily Hampshire Gazette

Essex County Herald

Vermont Watchman

OTHER

Court records of cases involving George Van Dyke and his various corporations.

Index

Page numbers in *italics* refer to illustrations.